Hazardous Waste Cost Control

COST ENGINEERING

A Series of Reference Books and Textbooks

Editor

KENNETH K. HUMPHREYS, Ph.D.

Consulting Engineer
Granite Falls, North Carolina

Executive Director, AACE International, 1971–1992

Additional Volumes in Preparation

Hazardous Waste Cost Control

edited by
Richard A. Selg

Westinghouse Savannah River Company
Aiken, South Carolina

Marcel Dekker, Inc.   **New York • Basel • Hong Kong**

Library of Congress Cataloging-in-Publication Data

Hazardous waste cost control / edited by Richard A. Selg
     p. cm. -- (Cost engineering ; 20)
    Includes bibliographical references and index.
    ISBN 0-8247-8891-5 (alk. paper)
    1. Hazardous wastes--Cost control.   I. Selg, Richard A.
      II. Series: Cost Engineering (Marcel Dekker, Inc.) ; 20.
TD1050.C67H38   1993
628.4'2'0681--dc20   93-17312
                                                CIP

The publisher offers discounts on this book when ordered in bulk quantities. For more information, write to Special Sales/Professional Marketing at the address below.

This book is printed on acid-free paper.

Marcel Dekker, Inc.
270 Madison Avenue, New York, New York   10016

Current printing (last digit):
10   9   8   7   6   5   4   3   2   1

PRINTED IN THE UNITED STATES OF AMERICA

Preface

Hazardous Waste Cost Control was written with project managers, estimators, cost engineers, schedulers, environmental engineers, and upper-level undergraduate and graduate students in mind.

This text presents a comprehensive approach to resolving such issues as cost growth, environmental regulation concerns that impact the schedule and cost of a project, static and dynamic baseline development, cost estimating, calculating contingency, risk and uncertainty analysis, innovative claims and dispute avoidance.

This timely resource emphasizes the following extremely important aspects of hazardous waste cost control by

- explaining which factors cause cost growth in hazardous waste cleanup projects and how to hypothesize on the drivers of this cost growth
- describing the key factors in successful estimating and scheduling of any hazardous waste project
- addresses the potential impacts that current environmental regulations and agency negotiations can have on a project—including identification of the regulations that will be applied to a project—in addition to suggesting how to evaluate these regulations and how to include their requirements as part of the project plan
- presenting the development of hazardous waste baselines that represent the planning process for both the assessment and remediation phases of environmental cleanup

- providing highly successful methods of contingency and risk analysis to assist decision makers in prioritizing and executing environmental projects that support a substantial reduction in cost and human health risk
- focusing on the reasons for claims and disputes in hazardous waste projects with emphasis on the innovative ways architects, engineers, contractors and owners can avoid or reduce the high costs associated with these claims

Industry-wide acceptance of these cost control principles and accelerated implementation of these concepts through executive management's commitment should pave the way for efficient planning and execution of environmental programs. This will ensure maximum returns for our environmental dollars.

Special acknowledgment is given to Dr. Kenneth Humphreys (Past Executive Director of AACE International) for his commendable service in reviewing this first edition. Our gratitude is also extended to AACE International for their support and sponsorship of this book.

As editor, I wish to express my appreciation to all the contributing authors for their exemplary effort to develop and prepare their individual chapters, tables, and appendixes. Special thanks to Dr. Rick Shangraw, Jr., his staff, and each of the contributing authors for assisting me with the final enhancement and update of this book.

If the reader considers the coverage in this book to be informative and reasonably complete, the presentation clear and concise, and the format pleasant to use, our major goals will have been achieved.

Richard A. Selg

Contents

Contributors

Donald J. Cass, CCE, is Principal and Project Controls and Construction Management Consultant of Cass & Associates in Los Angeles, California. Mr. Cass is a Fellow and Certified Cost Engineer, AACE International, an International Certified Cost Engineer, and a Fellow, Institute for Advancement of Engineering. He is a reviewer for the AACE Cost Engineering Magazine and a visiting lecturer at the University of Southern California in construction management. Mr. Cass has published dissertations on a variety of cost engineering subjects worldwide. He is a graduate of New Mexico State University with an MS in construction management pending from the University of Southern California.

Kenneth R. Cressman, CCE, is a Project Controls Manager for Jacobs Engineering Group, Denver, Colorado, with over 15 years of professional experience in cost estimating and cost engineering. He has performed cost analyses and developed cost estimating relationships for numerous environmental and hazardous waste projects. He is a graduate of Colorado State University with a B.S. in Biological Science. Mr. Cressman has taken graduate courses in finance, management accounting, macro- and micro-economics, and project management at the University of Denver. He was a lecturer in project management at the University of California–Irvine.

He is a Certified Cost Engineer, AACE International. He is also a member of Performance Management Association and the Order of the Engineer.

Gerald P. Klanac is a Regional Manager with Project Management Associates, Inc. (PMA), Ann Arbor, Michigan, where he is actively involved in all facets of construction management consulting including claims avoidance and labor productivity analysis. Prior to his tenure with PMA, he was a senior cost and schedule control engineer with Exxon Research and Engineering Company serving overseas in England on several refinery projects. Mr. Klanac holds an M.B.A. from the University of Chicago and a B.S. in Civil Engineering from the University of Dayton. Mr. Klanac is the author of numerous publications and has presented several papers throughout the United States.

Bruce A. Martin, CCE, is a Senior Cost Engineer with Black & Veatch, Overland Park, Kansas. Mr. Martin is a Certified Cost Engineer, AACE International. He has written articles on cost engineering for publication in the journals of AACE and PMI. He has served on the AACE Board of Directors in the offices of Technical Director, Vice President–Technical, Vice President–Regions, and Vice President–Finance. His work experience includes manufacturing, engineering and construction firms engaged in a variety of domestic and international environmental and industrial projects. Mr. Martin has presented several papers at national conferences throughout the United States.

Timothy C. McManus is a Senior Principal and Northeast Regional Manager with Project Management Associates, Inc. (PMA), in Boston, Massachusetts, where he is actively involved in a variety of projects in the hazardous waste, environmental, infrastructure, transportation, architectural, and industrial fields. Mr. McManus is a member of AACE International and the Water Environment Federation. He has lectured and presented numerous papers at conferences throughout the United States. He has taught at Harvard University's Graduate School of Design, and is currently lecturing at the Northeastern University Graduate School of Engineering. He is a graduate of Manhattan College with a B.S. in Civil Engineering and has studied Business Administration at the Seton Hall University.

Gui Ponce de Leon, Ph.D., P.E., has been the Managing Principal of Project Management Associates, Inc., based in Ann Arbor, Michigan, since

1971. He received his B.S. in Civil Engineering from the National University of Peru, and earned his M.S. and Ph.D. in Civil Engineering from Vanderbilt University and the University of Michigan, respectively. Over the last 21 years he has developed PMA from a scheduling firm into one of the country's fastest growing project management consulting firms. From 1973 to 1989 he also served as an adjunct professor in the University of Michigan's Graduate Construction Management Program, and is a registered Professional Engineer in Michigan, Florida, Arizona, and Louisiana. Dr. Ponce de Leon is a leader in claims and dispute mitigation and avoidance. He is nationally recognized in the evaluation of complex claims and has provided expert testimony in state courts, federal courts, and arbitration proceedings.

Richard A. Selg, CCE, is a graduate of the University of West Florida with a B.S. in Industrial Technology with over 21 years experience as a Senior Project Engineer, Project Manager, Project Director and Chief Cost Engineer in a wide variety of domestic and international environmental/hazardous waste and industrial projects. Most recently, he has been actively involved in program control and development at the Westinghouse Savannah River Company's Environmental Restoration Department, Aiken, South Carolina. Mr. Selg is an experienced trainer in project management, project controls, estimating, and human relations in the field of environmental/hazardous waste for the petrochemical, chemical, and nuclear industry. He is the author of numerous publications and has presented several papers at national and international conferences throughout the United States, Europe, and the Middle East. He is a Certified Cost Engineer, AACE International. Mr. Selg is presently serving as the AACE National Chairman of the Environmental Cost Control Committee. He is also a member of Project Management Institute and the Order of the Engineer.

R. F. Shangraw, Jr., Ph.D., is President of Project Performance Corporation (PPC), Sterling, Virginia, and a specialist in large-scale project and program planning. Dr. Shangraw has developed and implemented cost and schedule risk analysis systems, project benchmarking systems, and large-scale management systems in both the public and private sectors. Most recently, he has been a consultant to the U.S. Department of Energy's Office of Environmental Restoration and Waste Management on cost and schedule issues. Prior to founding Project Performance Corporation, Dr. Shangraw served as Vice President of Independent Project

Analysis, Inc., and held a joint faculty appointment in the schools of Public Administration and Engineering at Syracuse University. Dr. Shangraw received his Ph.D. in Technology and Information Management from the Maxwell School at Syracuse University. He is the author of numerous publications in project and program management.

Ronald G. Stillman, PE, CCE, is a Project Controls Manager for Roy F. Weston, Inc., West Chester, Pennsylvania, with over 18 years of project controls experience on diverse engineering and construction projects in the field of power, pharmaceutical, and the remediation of hazardous waste sites. Prior to his present position, Mr. Stillman was the Manager of Cost Engineering and Estimating for United Engineers & Constructors, Inc. He has lectured and presented numerous papers at National Conferences throughout the United States. He is a Certified Cost Engineer, AACE International, as well as a Professional Engineer (Colorado and Massachusetts). He is a graduate of Cornell University with a B.S. in Civil Engineering. Mr. Stillman is also a graduate of Northeastern University with an M.S. in Engineering Management.

Gary E. Thompson is a Staff Member at the Los Alamos National Laboratory, Los Alamos, New Mexico, for the Environmental Restoration and Waste Management Technical Support Office. He currently serves as an advisor to the DOE Office of Waste Management. He received his B.S. from the University of New Mexico. Mr. Thompson is also in the process of obtaining an M.B.A. He has worked in both the public and private sectors in program and project controls systems. The author of numerous publications, Mr. Thompson has presented several papers at national conferences throughout the United States.

William A. Zbitnoff, PE, CCE, is a Supervising Engineer with EMCON Northwest, Inc., Bothell, Washington, an Environmental Consulting Firm with 35 offices throughout the country. He has 18 years experience in hazardous/industrial waste management and petrochemical construction projects. He has worked on several hazardous waste site investigations and remedial engineering projects. Mr. Zbitnoff has also managed projects that have resulted from USEPA compliance orders, CERCLA National Priority Listing, and RCRA site closure actions. He continues to closely monitor environmental regulatory activities to determine their cost and schedule impact on site investigations and cleanups for several major projects. Mr. Zbitnoff has presented numerous papers at national conferences throughout the United States. He is a Certified Cost Engineer, AACE International, and a Professional Engineer.

Marc A. Zocher, CCE, is currently Western Operations Manager for Project Time & Cost, Inc., Albuquerque, New Mexico. He has spent seven years working in the DOE Complex as a Staff Member for Los Alamos National Laboratory. He has worked the last two years assisting the DOE Office of Waste Management in Washington, D.C. on matters of cost engineering applied to the overall DOE Program Management in Waste Management and Environmental Areas. Mr. Zocher has also held visiting faculty appointments at the University of New Mexico. He received his B.S. in Geology from New Mexico State University. He is the author of numerous publications and conducted several workshops on environmental cost estimating, activity-based costing (ABC), baselining, and project and performance control. He is a Certified Cost Engineer, AACE International.

Hazardous Waste Cost Control

1
Hazardous Waste Cost Control Management

Richard A. Selg
Westinghouse Savannah River Company, Aiken, South Carolina

1.1 INTRODUCTION

Total cost management techniques are vitally important in our fast-changing world of industry and commercial endeavors. Total cost management is the effective application of professional and technical expertise to plan and control resources, costs, profitability, and risk. Managers, engineers, and educators at government facilities, commercial industry, and universities around the world are discovering these benefits of total cost management in the field of hazardous waste and environmental restoration.

The United States' industry generates approximately 12 billion tons of solid waste annually, of which about 750 million tons are hazardous. Large amounts of energy, four to seven quardrillion BTUs (quads), are expended each year to produce and process raw materials that eventually become waste. Two additional quads are used annually in waste treatment and disposal operations. Recent estimates indicate that U.S. industry and public agencies spend nearly $20 billion per year for environmental control of waste.

It is no wonder that the cost of environmental restoration of hazardous waste is a subject that is both timely, revelant, and of interest to so many across the globe. It is important that the general public and readers of this text understand that we are managing the cost of hazardous waste restoration in all its dimensions through total cost management techniques such as those presented in this text.

In managing hazardous wastes, industry and government have their respective roles to play. An ever-increasing part of that role and the key to success are sound cost control practices and techniques.

This chapter will provide basic (generic) cost control concepts for managers and engineers in the field of environmental restoration and waste management and an excellent outline for upper-level undergraduate and graduate students in environmental engineering and sciences.

1.2 THE PURPOSE OF COST CONTROL

As defined by the American Association of Cost Engineers, cost control (project controls) is the application of procedures to monitor expenditures and performance against progress and manufacturing operations with projected completion to measure variances from authorized budgets and allow effective action to be taken to achieve minimal costs (and acceptable timetables).

The basis of an effective project control system is successful project execution. Although success is a relative term, it can be perceived as the timely completion of a project, every objective of which has been implemented to the owner's satisfaction. To this end, the contractor must establish trend-based information from which objective evaluations may be made. Project progress trends can be analyzed for various areas of work, and the level of performance as well as the efficiency of the work itself can be objectively measured. Performance measurement, obtained by various means, can provide for an impartial assessment of productivity and efficiency. If inefficiencies are found, this knowledge can be utilized in a constructive manner by project management to improve performance in applicable areas warranted by the analysis.

1.3 MANAGEMENT STRATEGY

There exist three major styles of management: (1) management by objective, (2) management by exception, and (3) management by crisis.

1.3.1 Management by Objective

Management by objective is the organization and control of program strategies whose primary goal is to achieve an end result, with the likelihood of success being dependent upon the satisfactory completion of incremental milestone objectives.

1.3.2 Management by Exception

Management by exception focuses on problem areas, the resolution of which results in the timely completion of activities considered problematic, and also includes management control of the critical path.

1.3.3 Management by Crisis

Management by crisis allows for the disruption of the normal flow of operations by concentrating on extinguishing the fires that seem to persist.

1.4 THEORY OF INTEGRATED PROJECT CONTROL

The project control system establishes guidelines for effective cost and schedule control. The basic principles consist of controlling elements of the project to a level of detail that can be monitored and controlled. The primary driver of the control system is the Critical Path Method (CPM) network; the activities in the network are sequenced to show the flow of the work to complete the project.

The project control system effectively performs the following functions:

- Develops strategic project plans to determine the timing of all major activities and to identify areas of critical concern. (Refer to Chapter 4 for more specific information on cost/schedule baseline development for hazardous waste projects. As regards areas of critical concern refer to Chapter 2 on environmental regulation impacts and agency negotiations and Chapter 3 on innovative claims and dispute avoidance.)
- Organizes project plans to establish the timing of all major components of the overall plan. (Refer to Chapter 4 for more specific information on cost/schedule baseline development for hazardous waste projects.)
- Establishes cost estimates to evaluate the relationship of budget appropriations to elements of the project plan. (Refer to Chapter 5, 6, and 7 for more specific information on estimating, cost drivers, and contingency estimating for hazardous waste projects.)
- Verifies that activity budgets are properly allocated within appropriate time periods for all labor, material, and equipment to be expended. (Refer to Chapter 4 for more specific information on budgets.)
- Determines critical paths to identify schedule priorities with analysis directed at resolution.
- Reports timely project status to management.

The project control system, mentioned in Section 1.4, is a dynamic process in which management direction is derived from the primary indicators of schedule status and cost performance.

1.5 MAJOR ELEMENTS OF PROJECT CONTROL

The major elements in developing effective project control methods are sequenced as follows:

1. Plan
2. Organize
3. Direct
4. Analyze
5. Control

1.5.1 Plan

Planning is the process of establishing major milestones and objectives for the project. This is generally accomplished at the request for the proposal stage. The two primary indicators of project progress are cost and schedule status. The development of *work packages* is a viable means by which to manage the project. Work package development usually consists of identifying work activities coupled with an understanding of the resource (labor, material, and equipment) requirements to perform each task.

1.5.2 Organize

Organizing the plan is simply developing a more complete understanding of what the plan entails. This effort commences at the time of contract award and includes determination of the project working budget. Information such as the identification of discrete work activities and their associated durations, relationship between tasks, constraints, costs, etc., results in the ability to perform cost/schedule calculations. Although the purpose of this chapter is not to identify the wide variety of project control methodologies that can be used, it is sufficient to state that information such as the calculation of early/late start and finish dates, scheduled float, criticality, as well as earned value information and cost performance results in an organized means by which to track cost and schedule progress. This is the developmental phase of the project; each work package is specifically identified, a value ranking (prioritization) is assigned, and the work package program is implemented.

1.5.3 Direct

In conjunction with organizing the development of resource and financial requirements for the project, the detailed aspects of project execution (design, engineering, procurement, manufacturing, installation, start-up, etc.) are implemented.

In effect, project management administrates the organized plan as described above.

1.5.4 Analyze

Once the detailed plans are developed and under way, the evaluation of actual progress against established targets can be initiated. (Refer to Chapter 4 on baseline development.) This process is fourfold:

1. Identify actual and potential problems.
2. Evaluate impacts of the problems.
3. Recommend solutions to problems.
4. Select corrective action.

1.5.5 Control

The project control team is generally regarded as an information gathering and dissemination service. To gain the due respect of the project team for the expert service that professional project control personnel can provide depends on their ability to accurately plan, organize, direct, and analyze potential and actual deviations to the planned budget (cost/schedule baseline). The root cause of a project manager's ability to control a project is simply *communication*. Through the analysis received by the project control personnel, the project managers can assess the impact of the project cost and schedule and direct the necessary positive action to minimize the overall impact to the project.

There are a variety of ways in which to depict project status, the culmination of which is the development of project status reports, from which project control strategies can be implemented by management. Most managers focus on what is going "wrong" with a project—this is a reasonable approach to project management (called *management by exception*) because cost/schedule overruns can usually be balanced against underruns, except in the case of critical path activities.

1.6 ASPECTS OF EXCELLENCE IN PROJECT CONTROL

Simply stated, *project control* as envisioned by industry today is a reporting procedure (or system) for alerting project management of anticipated directions in the project plan concerning time and money. In the case of adverse trends, it is an "early warning system" that allows the *project team* sufficient time to take corrective action in a manner that avoids and/ or minimizes adverse consequences (sudden and unexpected surprises) to the proposed schedule and costs.

The attributes of good project control, stated in their order of importance, are: (1) status reports and analysis, (2) forecasts, and (3) implementation of procedures.

1.6.1 Status Reports and Analysis

Status reports should contain the following: (1) total probable project cost (forecast—refer to Section 1.6.2), (2) highlight of variances (to the budget), (3) clear and concise information, and (4) a uniform format.

1.6.2 Forecasting Costs and Schedules

Forecasts that are timely and performed routinely (at least once a month; more often if deemed critical by management). To be effective, cost control *must* be based on the project forecast, not on the authorized budget, and *must* be an integration of schedule and cost data, which should be based on the following important factors:

1. Representation of engineering and construction progress and efficiencies
2. Procurement performances
3. Resource utilization
4. Inflationary influences
5. Other factors contributing to time and money deviations

1.6.3 Implementation of Procedures

The following rules should apply to the implementation of procedures: (1) the implementation of the procedures should provide opportunities to ensure hard dollar savings, (2) the procedure development should require minimal investment, and (3) the use of the procedure should result in a return of investment for the procedure development.

1.6.4 Integration of the Cost and Schedule

The integration of cost/schedule information is incomplete without a defined method of communicating this information to managers and responsible team members for analysis, decision making, and action. Other critical items to be considered are as follows:

1. Engineering and drawing status
2. Procurement status and materials management
3. Scope development and scope changes
4. Progress trends
5. Actual and forecasted quantities
6. Schedule activity relationships and logic, etc.

In summary, the dedicated intent and purpose of a project controls group should, therefore, be to monitor, analyze, and correlate cost and schedule data into a management information system with regularity, consistency, and accuracy.

1.7 GUIDELINES FOR HAZARDOUS WASTE COST CONTROL MANAGEMENT

There is an increasing need for a project waste management plan. A waste management plan can provide internal guidance and cost control measures, as well as provide evidence of a company compliance program to an external agency such as the EPA. An initial set of guidelines should be used to reflect accurately the costs of waste disposal in the project estimate. (Refer to Chapters 5, 6, and 7 for more specific information on hazardous waste estimates.) These guidelines can then be converted into a project plan for waste management. The primary objectives of a waste management plan should be:

- Cost control of fees, permits, labor, and equipment to manage the project waste
- Control of procedures, resulting in orderly sorting, storing, and disposal of the waste material
- Elimination of fines, investigations, and regulatory disruption resulting from improper waste management
- Job site safety: Coordination with the safety program concerning toxic substances and their proper disposal

The project management plan should consist of the following elements (Also refer to Figures 1.1 and 1.2):

WASTE MANAGEMENT SURVEY FORM *(For use before construction begins)*

Project/Site Waste Stream Report

Project Name: _____   **Bid Date:** _____

                                                                    Date: _____

Project Location:_____   **By:** _____

If during specification review, potential recycle was no, explain:

_____

Plan for Recycling:

_____

Plan for source Reduction:

_____

Remarks:

_____

(a)

WASTE MANAGEMENT DISPOSAL RECORD *(For Use During Construction)*

PROJECT: DATE:

LOCATION:

MATERIAL

CLASSIFICATION

QUANTITY

LOCATION DISPOSED

METHOD OF DISPOSAL(Landfill, recycle, exchange, etc)
COST:
   LABOR:
   EQUIPMENT:
   TIPPING FEE:

(b)

Figure 1.1   (a) Waste management survey form and (b) waste management disposal record.

Subcontract Agreements

General contractors presently are attempting to move more of the clean-up responsibilities to their subcontractors. Disposal in the commercial sector is still being paid for by the general contractors as a general condition item, but in residential construction the trend is to put the burden of both the clean-up and disposal on the subcontractor. The following subcontract agreement (Example 1) demonstrates this recent trend to stricter subcontract clauses concerning cleanup and disposal. The next step in this process will be to define their responsibilities in connection to the broader picture of waste management and not just waste disposal. Example 2 adds additional language that could be included in special/supplementary conditions of the subcontract. Thorough education of subcontractors concerning separation techniques, recycling techniques, and source reduction techniques of construction waste materials will enable both general contractor and subcontractor to effectively reduce the cost of solid waste being produced in the field.

Example 1——Typical Subcontract Agreement
(Commercial/Industrial Projects, Larger Scale Projects)

"Q. **CLEAN-UP** - During the course of construction, SUBCONTRACTOR shall remove waste materials from the site sufficiently as is necessary to maintain the premises in a clean and orderly condition. Upon completion of the work under this Agreement, SUBCONTRACTOR shall remove from the site all temporary structures, debris, and waste incident to its inspection and clean all surfaces, fixtures, equipment, etc relative to the performance of this Agreement. If SUBCONTRACTOR fail to perform a clean-up function within two days after notification from CONTRACTOR to do so, CONTRACTOR may proceed with that function as it judges necessary and in the manner it may deem expedient, and the cost thereof shall be charged to the SUBCONTRACTOR and deducted from monies due under this Agreement.

.... supplementary conditions to standard clauses above ....

C. Qualifications

1. ....

2. All stocking of material and debris removal to be performed during the off-hours. Hoists are for personnel only between 7:00 a.m. and 3:30 p.m. The subcontractor shall bear the cost of after-hour hoist operator, who shall be billed at $ 65.00/hour."

Example 2——Waste Management Supplemental Provisions to Standard Subcontract Agreement
(Commercial/Industrial Projects, Larger Scale Projects)

C. Qualifications

1. ....

2. ....

3. **WASTE MANAGEMENT PLAN** - All subcontractors will be required to strictly adhere to the CONTRACTORS' Waste Management Plan and to file with the CONTRACTORS' office and obtain approval of, his own waste management plan which will be equal to or more thorough in source reduction and recycling for the waste products generated by this SUBCONTRACTOR prior to beginning work. During the clean up of SUBCONTRACTORS' waste, the SUBCONTRACTOR will work with all parties to provide and be part of a comprehensive program of waste management. Any additional costs incurred by the CONTRACTOR due to lack of performance and strict adherence to the SUBCONTRACTORS' waste management plan will be charged to the SUBCONTRACTOR and deducted from monies due him/her under this Agreement.

Figure 1.2 Typical subcontract agreement and waste management supplemental provisions.

- Survey of project wastes, including: demolition material, material waste, packaging estimates, and potentially toxic or hazardous wastes
- Analysis of cost implications of waste management for the specific project;
- Analysis of the plan for the collection and disposal of the project waste, including: (1) a collection system for each type of job site, including point of storage, separation, and pickup; (2) a plan for subcontractor compliance to the waste removal plan, indicating the materials of waste and how they are committed to the location of disposal from the responsible party; and (3) a system of recording the removal of waste from the job site, indicating the type of waste, the amount of waste, and the disposal location of the waste.
- Reports kept in the project records involving waste and waste disposal;
- Coordination with the project safety program concerning disposal of toxic chemicals included in the project Material Safety Data Sheets (MSDS).

In addition to the project waste management program, the various construction firms should have an overall waste management program, similar to a company safety program, to provide a uniform direction and awareness to construction waste management. A company waste management program should have the following elements:

1. A statement of purpose and commitment to the program from upper management, indicating that the purpose of the program is to dispose of construction waste in compliance with current and future regulations, and to keep the amount of waste to a minimum within the cost constraints. (Refer to Chapter 2 for more specific information about environmental and hazardous waste regulations.)
2. A specific procedure, including the survey, estimating procedures, disposal guidelines, and record keeping.
3. A training program for project managers and superintendents that involves the importance of proper disposal of hazardous material, the methods and procedures for disposing the hazardous material, and the necessary record keeping.
4. A specific individual in charge of the program to administer its use and to keep the company, contractor, and customer abreast of the changes in the regulations. This person would be the prime liaison with the regulatory agencies such as the municipal/county health districts, state departments of ecology, and the EPA. (Refer to Chapter 2 and Appendix C for environmental and hazardous waste regulations and agencies.)

1.8 CONCLUSION

The ability to control costs and schedules is essential for corporate success in the business world of today's tough competition and economy. Every project manager on a hazardous waste capital project would like to make the following entry to the final project report: "Project completed on budget and on schedule with an excellent start-up!" Indeed, the success of hazardous waste construction projects is judged in terms of achieving the three interdependent project management objectives of quality, cost, and schedule. Quality in terms of budgeting, planning, and project management has a direct impact on the success of project cost control. The concept that *time is money* also has a direct bearing on the project's final results, necessitating comprehensive planning, monitoring, and follow-up to avoid costly delays of the project schedule.

The *Cost Engineer* is the focal point for the project control system. The objective is to minimize the impact on the cost and schedule by focusing management's attention on potential cost and schedule trouble spots in time for corrective action to be taken.

A *Hazardous Waste Management Program* should aim to be cost effective without disrupting the building process. It should complement the other control systems of cost control, quality control, schedule control, and safety. The program must be pro-active, such as a safety program, with training and monitoring as the keys to success.

REFERENCES

Clark, F. D., and A. B. Lorenzoni (1978). *Applied Cost Engineering*, Marcel Dekker, New York.

Lutz, D. D., and T. J. Buechler (1992). Evaluating the Cost of Remedial Action. *AACE Transactions*, Paper No. H-1.

Martin, B. A. (1992). Aspects of Cost Control, *Cost Engineering* 34 (6):19–23.

McMullan, L. E. (1991). Cost Control—The Tricks and Traps. *AACE Transactions*. Paper No. O-5.

Rivera, A. L. (1990). Challenges in Estimating Waste Confinement Costs. *AACE Transactions*. Paper No. J-5.

Johnson, H., and W. R. Mincks (1992). Waste Management for the Construction Manager. *AACE Transactions,* Paper No. J-5.

2

Environmental Regulation Impacts and Agency Negotiations

William A. Zbitnoff
EMCON Northwest, Bothell, Washington

2.1 INTRODUCTION

This chapter will address the potential impacts that current environmental regulations and agency negotiations can have on a project. It will identify several regulations that are likely to be applied to a project and will make suggestions on how to evaluate these regulations and include their requirements as part of the project plan.

Perhaps the most important step in a project, given our current regulatory environment, is to evaluate the environmental conditions of the project, including the history of site property. This is very important during property acquisition or selection of a project location. The majority of property transfers and other transactions are now contingent on conducting an environmental site assessment of a property or facility.

Potential project cost and schedule impacts from environmental regulations can be reviewed in a logical sequence by addressing a project in the following three phases: (1) initial property or facility acquisition, (2) site preparation and demolition, and (3) facility construction. We will look at each of these project phases and identify areas of concern and the potential impact of regulations. In addition, we will review how to negotiate and manage a hazardous waste investigation and remedial action.

2.2 ENVIRONMENTAL REGULATIONS

Maintaining a general understanding of the environmental regulations and policies in effect is the first step toward incorporating their impacts into the project plan properly. It is important to acknowledge that many environmental clean-up activities have been undertaken because they are mandated by federal, state, and local environmental regulations. Therefore, a project manager must be aware of the regulations that may impact a project and the regulatory agencies that are administering those programs.

There are numerous federal laws that may impact a project. Some of these are:

- Clean Air Act (EPA)—regulates emission of hazardous air pollutants.
- Clean Water Act (EPA)—regulates discharge of hazardous pollutants into the nation's surface waters.
- Comprehensive Environmental Response, Compensation, and Liability Act (EPA)—provides for clean up of hazardous waste sites.
- Occupational Safety and Health Act (U.S. Occupational Safety and Health Administration)—regulates hazards in the workplace, including worker exposure to hazardous substances.
- Resource Conservation and Recovery Act (EPA)—regulates hazardous waste generation, storage, transportation, treatment, and disposal.
- Safe Drinking Water Act (EPA)—regulates contaminant levels in drinking water.
- Toxic Substance Control Act (EPA)—regulates the manufacture, use, and disposal of chemical substances.

In addition to these laws, there are several other federal, state, and local environmental laws and regulations that can impact a project. Because the regulations and enforcement responsibilities vary from state to state and because the regulations evolve at a very rapid rate, it is beyond the scope of this or any other single text to provide a complete up-to-date reference of all environmental regulations that may be applied to a given project. We will concentrate on the two primary federal regulations that will have the most impact on environmental projects: the Resource Conservation and Recovery Act (RCRA) and the Comprehensive Environmental Response, Compensation, and Liability Act (CERCLA). We recommend that if you have questions regarding regulatory compliance that you contact your environmental or legal council. We have also provided a listing of each of the 10 EPA regions, primary state regulatory agencies, and other informational contacts in Appendix C for your reference.

RCRA is the primary legislation concerning hazardous waste management and is designed to provide "cradle-to-grave" control of hazardous waste. RCRA focuses on facilities or sites that are currently dealing with hazardous wastes including possible storage of hazardous waste from past practices. CERCLA addresses the clean up of problems of past practices and immediate response to environmental emergencies. The majority of the monies currently being spent on environmental management is in response to the requirements of these regulations. The following information provides a general introduction to each regulation and identifies the regulatory actions that need to be considered.

2.2.1 Resource Conservation and Recovery Act

RCRA was initially passed in 1976 as an amendment to the Solid Waste Disposal Act to provide legislation for controlling hazardous waste management. This act was amended in 1984 with the Hazardous and Solid Waste Amendments (HSWA). RCRA is divided into three subsections: Subtitle C—hazardous waste, Subtitle D—solid waste, and Subtitle I—underground storage tanks. Subtitle C, the section that is most applicable to our needs, regulates hazardous material according to responsibility such as generating and transporting, or treatment, storage, and disposal. Each of these responsibilities has the potential to impact a project.

The most important aspect of the RCRA process is to determine if a material is a hazardous waste. There are two criteria that are used to evaluate if a material is a hazardous waste: (1) the material must be a solid waste, and (2) the solid waste must be considered hazardous. Part 40 of the Code of Federal Regulations defines a solid waste as any liquid, solid, semisolid, or contained gas that is discarded or stored prior to discarding. This definition clearly shows that the term *solid waste* is used for regulatory purposes and does not describe the physical characteristics of discarded materials.

After a material is identified as a solid waste, it must be evaluated to determine if it is hazardous. A solid waste is considered to be hazardous if it is listed in 40 CFR 261 or exhibits one of the following four hazard characteristics: ignitability, reactivity, corrosivity, or toxicity. The specific testing criteria for evaluating the four hazardous characteristics are defined in 40 CFR 261. There are some wastes, such as domestic sewage, irrigation return flows, special nuclear or nuclear byproducts, mining overburden, and other wastes that have been specifically excluded from Subtitle C. Determining if a material is a hazardous waste can be quite complex. It is one of the most important steps in evaluating how a material is handled during a clean-up project.

In addition to complying with RCRA regulations for the disposal of hazardous wastes, there are numerous other parts of this law that impact environmental engineering and hazardous waste management projects. We have only provided a basic introduction to the law and to those activities that may impact a project that is *not* a major hazardous waste clean-up. If you are starting a RCRA corrective action, closure of a RCRA impoundment, or other type of RCRA cleanup, we strongly recommend that you closely study one of the many texts that have been specifically written on RCRA and gain assistance from a professional who has completed similar RCRA projects.

2.2.2 Comprehensive Environmental Response, Compensation, and Liability Act

CERCLA, commonly referred to as "Superfund," is the other primary regulatory act that impacts many of the environmental management and hazardous waste projects. The law was enacted in 1980 to give the federal government the authority to clean up hazardous wastes and respond to releases of contaminants that pose a threat to human health and the environment. This law was amended in 1986 as the Superfund Amendments and Reauthorization Act (SARA). One of the major provisions in CERCLA as amended by SARA is the opportunity for the government to conduct site cleanups and determine the potential responsible parties (PRPs) that are liable for paying the costs of the clean-up activities.

The basic procedures to be conducted under Superfund are defined in the National Oil and Hazardous Substances Pollution Contingency Plan (NCP). The NCP establishes steps that the federal government must follow in responding to the release or threat of release of hazardous substances. These procedures include both immediate actions known as *response actions* and long-term cleanups known as *remedial-response actions*. The Superfund process begins upon discovery of a potentially hazardous site. The situation is reviewed to determine if the conditions require emergency actions or if a long-term action is more appropriate. Removal responses typically are completed in less than 1 year, with remedial responses being extended programs consisting of many steps that may take several years to complete.

The first steps of the remedial response process are site discovery and preliminary evaluation. Any federal, state, or local government agency may identify a site for future evaluation. The federal government then conducts a preliminary investigation to determine the extent of the problem and provide information that will be used to evaluate the priority for conducting remedial actions. Upon completion of this step it may be established that there is no threat or potential threat to human health and the environment and the site may not require any additional remedial action. The information from the preliminary assessment is used to calculate a hazard ranking system score to determine the priority for conducting a site cleanup. Depending on the hazard ranking system score a site may be placed on the National Priorities List (NPL) as a site that poses significant threat human health and the environment and must be remediated.

If a site is placed on the NPL, cleanup actions are evaluated and selected by a remedial investigation and feasibility study (RI/FS) process that is specifically defined in federal guidance documents. This process is an integrated approach that characterizes the site conditions, assesses the potential human health and environmental risk resulting from site hazards, identifies numerous cleanup and treatment technologies, and reviews the applicability of these technologies to meet long-term remediation objectives. An RI/FS and site cleanup may be conducted by the PRPs or by the federal or state government and includes a high level of public participation. The recommended remedial action must meet CERCLA requirements but must also comply with other state and federal applicable or relevant and appropriate requirements (ARARs). Upon acceptance of an RI/FS by the EPA a remedial action is recommended and documented in a Record of Decision (ROD); the remedial action is then conducted in accordance with the ROD.

The process described above is a very brief summary of the numerous and sometimes time-consuming steps that are required for remediation of a NPL site. The importance of understanding this process prior to conducting a CERCLA cleanup can not be overstated. There are numerous pitfalls that can delay a project and cost thousands of dollars. It is important to know that if you are working at a site that becomes a NPL site that there are several specific steps that must be followed in order to comply with the federal regulations.

2.3   PROJECT IMPACTS BY PROJECT PHASE

2.3.1 Initial Property or Facility Acquisition

There are several environmental regulations that are of concern to a project and should be evaluated before property is purchased or designated for a project. Many of these regulations have to do with existing site conditions or possible contamination that may have originated at the potential site. We will evaluate potential regulations and identify impacts that should be included in the project plan.

Due Diligence

The federal government, along with the majority of states in the country, have established hazardous waste regulations that are designed to protect human health and the environment. Many of these regulations have established extensive lists of contaminants and contaminant concentrations that are considered hazardous to human health and the environment. The presence of these contaminants causes concern to the environmental regulatory agencies and may necessitate investigation and cleanup of a site. Current hazardous waste regulations have established that the cost for cleanup must be paid for by both the present and past owners of the site, as well as any parties that generated contaminants or transported them to the site.

There are few defenses that will allow a property owner to avoid paying for cleanup of the site. These expenses can easily cost millions of dollars. One of the most common defenses to lessen or defer investigation and cleanup cost is to qualify oneself as an innocent purchaser. To meet this defense the purchaser must demonstrate that the contamination occurred prior to his purchase of the site and that he "did not know and had no reason to know" that the site was contaminated. This burden of proof lies with the purchaser of the property. Therefore, when purchasing property or facilities the purchaser must be able to demonstrate that he used "due diligence" to investigate for the presence of hazardous constituents; if he is unable to demonstrate this he will be responsible for cleanup costs.

Based on the potential cost and liability of these statutes, property owners, banks, insurance companies, and other lending institutions require environmental site audits to be conducted prior to the transfer of ownership for property or facilities. These audits or assessments vary in cost from $2,000 to $3,000 for small sites that do not require any sampling and analysis, to $100,000 for a large industrial facility that may have numerous potential problems. Typically, audits can be completed within 3 weeks for a small facility or can take several months depending on the complexity of the site and the level of contamination.

Unfortunately there are no rules of thumb that can be universally applied to all sites. The cost and duration of a site audit is dependent on the site history, surrounding areas, and the state and federal regulatory climate.

Each of these factors can easily modify the scope of work. Another significant factor is the amount of time that is allowed for the analysis. Typically, analytical results require 30 to 45 days. This duration is from the time the samples are received until results are available. Another 7 to 21 days need to be added for verification and review of the results. This 1 to 2 months can have a significant impact on a project schedule if this activity has not been included in the project plan. It should be noted that this analysis can be expedited but the costs will be increased by as much as 100% depending on the laboratory schedule and the number of samples analyzed.

There are no environmental regulations that require an owner to conduct a site audit; however, it is recommended that this type of investigation be incorporated into any project plan that will involve newly purchased property or facilities. Site audits should also be conducted for projects at locations that you already own but of whose current or past environmental status you are unaware. If the study has been done correctly, the owner should be well aware of any problems that could impact an upcoming project. Typically this type of study will also identify potential problems that may not be in compliance with other environmental regulations. This process is quick and inexpensive, and can save major unforeseen cost and schedule impacts.

Clean Water Act Section 404 (Wetlands)

In addition to "due diligence" site audits, wetlands are a major topic of conversation and the source of numerous delays to projects throughout the country. Section 404 of the Clean Water Act prohibits the discharge of dredged or fill material into the nation's waters, including wetlands, without a permit from the U.S. Army Corps of Engineers (COE). Depending on the specific situation, this permit may be relatively easy to obtain in a period of 60 to 90 days with limited costs. However, it may not be possible to permit the project, or the permit process may require major modifications that are very costly and could take several months to complete.

Obviously, this regulation can strongly impact potential projects. The wetlands regulations are currently in a state of flux with the environmental activists on one side, industry and developers on the other side, and the regulators caught in the middle. As mentioned previously, this regulation is rather specific on the need for obtaining a permit. The problem with the regulations occurs with the definition of the terms of "nation's waters" and "wetlands." According to the COE definition, wetlands are:

. . . those areas that are inundated or saturated by surface or ground water at a frequency and duration sufficient to support, and that under normal circumstances do support, a prevalence of vegetation typically adapted for life in saturated soil conditions. Wetlands generally include marshes, swamps, bogs, and similar areas.

The interpretation of this definition can include a wide variety of areas. In Alaska, this includes the majority of the state. The federal government is currently seeking a "no net loss of wetlands" policy that will require an owner to protect any wetlands, create a new wetlands of equal or greater area, or possibly rejuvenate a wetlands that has been damaged in the past.

These regulations have impacted several projects currently in progress in western Washington. There are three types of permit application situations: (1) the permit is not challenged by anyone, (2) the permit is only challenged by neighbors and other local private organizations, and (3) the permit is challenged by state and federal agencies such as fish and wildlife protection agencies. In the first case, the permit will follow standard procedures and be issued in 60 to 90 days with minimum cost impacts. This process should be included in the project plan with minimum impact. The second case is slightly more complex. In the Pacific Northwest, the COE will issue a permit and leave resolution of the issues to the individuals and, possibly, the court system. If local groups are strongly opposed to the project, they can obtain an injunction from a local court that can stop a project for an indefinite length of time. These types of disputes are hard to predict and are dependent on a wide variety of circumstances. You, your public relations department, and your legal counsel are the best judge of the impact of this type of action.

The third case will always result in some delay of a permit. The COE will not issue a 404 permit until this type of concern is resolved to the questioning agency's satisfaction. In some cases this can be addressed quickly with simple modifications to a filling sequence or the timing or certain site work operations. In other cases, a fish or bird habitat may be deemed to be in danger and the project may not be allowed without major modifications. Wetlands regulations are becoming more restrictive every year. It is important to review site conditions carefully as a first step toward determining the feasibility of a project.

For example, a client had a project delayed for two construction seasons while a new wetlands was constructed to replace an area that was filled to allow for the expansion of the facility. This activity cost several million dollars and had a significant impact on the financial payback of the project. In this case, the project was still feasible; however, there are several other projects that have not been completed as the result of such an action.

Other Regulations

There are several other hazardous waste regulations that can impact a project from the beginning stages through final plant operation. Most of these regulations deal with obtaining operating permits under regulations such as RCRA. [Facilities that are governed by these regulations are identified Code of Federal Regulations.] The RCRA process is complex and deals primarily with long-term operation of facilities that handle or produce hazardous material or waste. This topic is beyond the scope of this text. If you are considering constructing a facility that will handle or produce hazardous wastes we recommend that you work closely with your environmental and legal council.

2.3.2 Site Preparation

There is a wide variety of federal and state hazardous waste regulations that may apply to any given site during the construction phase of a project. The primary federal regulations are CERCLA/SARA and RCRA that dictate the type of contaminants that are hazardous, the levels of contamination that require cleanup actions to be initiated, and establish cleanup or disposal methods that are acceptable. In addition to the federal regulations, many states have developed regulations that are as restrictive or even more restrictive than federal regulations.

State and Federal Hazardous Waste Regulations

Because the cleanup or action levels for given contaminants vary from state to state, we have not listed specific cleanup levels. It is recommended that you contact the EPA or local state agency to obtain the most recent information; they will be able to suggest possible actions that may be required to clean up the contaminants at your site. Fortuna and Lennett provide an interpretation of the hazardous waste definition stated in volume 40 of the Code of Federal Regulations (40 CFR) that is helpful in understanding if a material is a hazardous waste. This definition is as follows:

> A waste is considered hazardous if it falls within any of the four categories listed below and it does not qualify for any of the exemptions or exclusions stated in 40 CFR 261.
>
> 1. EPA may list a waste, usually from a specific production process, as hazardous, based principally upon the presence of specific hazardous constituents in the waste or because the water consistently exhibits one or more characteristics of a hazardous waste (40 CFR Part 261 Subpart D). EPA may also list a product as hazardous waste if it is discarded in a pure or off-specification form and contains specific hazardous constituents.

2. Those solid wastes and waste generation processes that have not been specifically listed by EPA may nevertheless be identified as hazardous solely on the basis that they exhibit one or more of the four characteristics of hazardous waste irrespective of the manufacturing process from which it is generated. The four characteristics are: ignitability (I), corrosivity (C), reactivity (R), or toxicity (EP). Toxicity (EP) means the ability or tendency to leach certain constituents via a specific extraction procedure (40 CFR Part 261 Subpart C).

3. It is a mixture of a listed hazardous waste and any other material, or is a mixture of a characteristic waste and any other material, provided the mixture still exhibits characteristic (40 CFR 261.3 (a) (iii) (iv)).

4. It is a residue that is "derived from" the treatment, storage, or disposal of a listed waste (40 CFR 261.3 (c)), such as incinerator ash.

These regulations frequently impact a project during the site preparation phase of a project. Any soils or demolition debris that are being removed from the site must be disposed of properly. If this material is considered hazardous by either state or federal regulations the disposal costs can increase dramatically. For example, soil removed from a site may be used as fill at another location. There could be less excavation and transportation costs and disposal may be free. If this same material is deemed a hazardous waste, the cost of disposal at a hazardous waste landfill can easily be as much at $300 per cubic yard, not including sampling and analysis costs to determine the level of contamination. Again, evaluating site conditions prior to beginning site work will identify hazardous material conditions so that they can be included in the project budget.

If there is a significant amount of contamination, or if the types of contaminants are extremely dangerous to human health or the environment, it is highly likely that either the state environmental agency or the EPA will become involved with the project. At this point the project takes on a new identity as a hazardous waste site and becomes much more complex. For example, a client had a site in which the initial site geotechnical investigation identified the presence of several heavy metals in the soils as being above state and federal maximum contamination levels. Construction was suspended until a hazardous waste remedial investigation and feasibility study could be completed and a cleanup action plan been approved for the site. This process may take a year or longer to complete and cost between fifty and several hundred thousand dollars depending on the level of contamination. These unplanned costs and schedule delays greatly impact the initial project plan and economics.

The basic message in this section again is to gain as much information about any given site at the very initial phase of a project. If there have been underground tanks, transformers, solvent cleaning activities, uncontrolled disposal practices, or numerous other activities at a site, proceed with caution and have a very flexible project schedule and budget. If a site becomes a state or federal cleanup site there are several activities that will have to be conducted. These activities are summarized later in this chapter. Unfortunately it is very difficult to predict accurately the overall cost and schedule impact that discovery of hazardous waste will have on a project. Several additional tools are provided in this text to assist with the development of a project plan.

National Pollutant Discharge Elimination System (NPDES)

The EPA recently passed NPDES regulations that are designed to control surface water runoff. Of specific interest to the construction industry is the section of the regulations concerning construction sites that are greater than 5 acres in size. In accordance with the latest regulations any construction activity that disturbs more than 5 acres of total land area requires an application for an NPDES permit. This permit application will be submitted to either the state or federal agency depending on which is the lead agency in that state.

This permit submittal is another of the environmental regulations that must be incorporated in the project planning process. The permit does not include any sampling and analysis data and should be relatively straightforward. However, depending on the nature of the receiving water it is entirely possible that the project could be delayed until a dry time of year. This change could have significant impact on a project schedule.

Other Regulations

Besides the primary regulations of concern there are requirements that may have been established as part of the 404 permit. Underground storage tanks (USTs) are another area of potential concern. There are currently regulations for installing USTs that must be followed regarding tank construction and long-term monitoring during operation of the facility. These regulations focus more on the design, operation, and maintenance of USTs and should not be a significant concern during a typical construction project. These are easily dealt with and must be incorporated into the project plan.

2.3.3 Facility Construction

During the remaining construction phases of a project environmental regulations should have limited impact with the exception of the phases of startup and operation. During these final stages of a project it is very important to have the proper permits such as NPDES or air quality permit in place prior to initiating any discharge. Another area of compliance is Superfund Amendments and Reauthorization Act Title III, which deals with public right-to-know information. This regulation requires notifying local health and safety organizations such as the fire department of hazardous materials that are on site and maintaining records and other information regarding these materials. Most construction organizations and owners are familiar with this act and the coordination of material safety and data sheets. This is an ongoing process and should not impact any given project.

2.4 MANAGEMENT OF HAZARDOUS WASTE PROJECTS

If a project becomes a state or federal hazardous waste project through the discovery of hazardous waste or contaminated soil or ground water, there are several steps that must be incorporated into the project plan. The specific steps are dependent on whether the site is being managed by the EPA or by a local state agency; however, the basis steps are the same.

- Collect initial site background information.
- Develop a plan for conducting a site investigation.
- Obtain regulatory agency approval of the investigation plan.
- Conduct site investigation.
- Compile results and review with regulatory agency.
- Prepare remedial action feasibility study.
- Obtain both agency approval for cleanup action.
- Conduct site cleanup.
- Conduct sampling and obtain agency sign off on project.

2.4.1 Agency Involvement

The number of steps listed above indicates that this process can be lengthy and expensive. A cost of several hundred thousand dollars and a schedule of 1 to 2 years is not uncommon; however, it is not correct to assume that this is the case for every hazardous waste investigation and cleanup. Several of these steps can be combined through an interactive relationship with the governing regulatory agency. Properly handling agency interface and negotiation can save significant costs and reduce the over-project duration.

Typically a regulatory agency will request a minimum of 30 days to review and approve any document submitted during a hazardous waste project. These reviews can have a definite impact on a project schedule with the submittal of both draft and final documents for each step of the process. However, it is possible to reduce this time if a good relationship is established with the agencies and they are brought into the planning process early on in the project. It is also important to remember that all projects and contaminants are not treated the same. If the project involves contaminants that are not exceptionally hazardous and the site is in an area with minimum impact to human health and the environment, it is possible that the project will be placed on a faster track. This can be accomplished by developing an understanding of the regulators' specific areas of concern early on in the project and by developing a project plan to meet those concerns. Quite often it is beneficial to have preliminary meetings with the regulatory agencies to discuss the project and make sure that all parties involved have a clear understanding of the project. This may be more difficult if more than one PRP has been identified and there is an adversarial relationship between the PRPs. It is better for all involved if a good working relationship can be developed with the regulatory agencies.

2.4.2 Project Strategy

One of the key steps to remember is that the primary purpose of the investigation is to gain enough information to prove that a site is not contaminated, or to identify a cleanup process that can be applied to the site. It is important to remember this strategy when developing an investigation plan. It is also important to remember that most hazardous waste projects will also need to be accepted by the public so the investigation that is conducted should be one that the public will believe is representative of the conditions at that site. Gather data that will be needed to design a cleanup strategy; do not just go through the investigation process hoping not to find any contamination. A well-planned and negotiated investigation will result in shorter total project schedule and will be more cost effective in gathering field data.

Work with the regulatory agency to determine exactly what their concerns are with the site. If they have only one area of concern, be sure to address that area and perhaps thus avoid the collection of unneeded information. Good relations with the agencies are not always easily attained. In fact, there are often situations when agencies appear to be making unreasonable demands that will greatly increase the cost of the project. It is important to educate the regulator about your concerns and reach a mutual understanding of the project requirements without alienating the regulatory agencies involved. Occasionally the project team will not agree with the regulatory requirements, but remember state and federal agencies win more arguments than they lose because they are protecting human health and the environment. Choose your battles wisely; select ones that make sense and that you can win and concede the others. Doing so will save time and money even though it may be very hard to accept.

When conducting the cleanup be sure to gain agency approval throughout the process. There is nothing worse than thinking that you have completed a project only to find that regulators did not understand what you were doing and want the work to be redone.

Overall there are a growing number of environmental regulations that will impact your existing and upcoming projects. Although a project manager may not be able to predict where the regulations will be next year, it is likely that they will be equal or more restrictive than the current regulations. It is important to become aware of the regulations that currently exist and to incorporate them into a project plan. Ignoring the potential impact of environmental regulations can be very costly and time consuming.

REFERENCES

Fortuna, Richard C., and David J. Lennett (1987). *Hazardous Waste Regulation: The New Era, An Analysis and Guide to RCRA and the 1984 Amendments*, 27, McGraw-Hill, New York.

National Pollutant Discharge Elimination System (NPDES) Permit Application Requirements for Storm Water Dischargers, (1990), EPA.

Soong, S. C., Elliot, J. F., and G. Goldfarb (1992). *Toxics Program Commentary*, Specialty Technical Publishers, Vancouver, Canada.

Requirements for Wetland Fill Projects (1989), Information paper, U.S. Army Corps of Engineers.

Freeman, Harry M. (1989). *Standard Handbook of Hazardous Waste Treatment and Disposal*, McGraw-Hill, New York.

Wagner, T. (1988). *The Complete Handbook of Hazardous Waste Regulation*, Perry-Wagner, Falls Church, Virginia.

3
Innovative Claims and Disputes Avoidance

Gui Ponce de Leon, Timothy C. McManus and
Gerald P. Klanac
Project Management Associates, Inc., Boston, Massachusetts and Ann Arbor, Michigan

3.1 INTRODUCTION

As the early 1990s unfold, construction investment for environmental projects is running at nearly $20 billion annually across the United States. If these projects experience the normal amount of disputes and claims, (i.e., roughly 15% to 25% of the contract values), there is between $3 to $5 billion of annual investment *at risk* as a result of claims and disputes. These estimated figures do not even reflect the millions of dollars of human capital spent by architects, engineers, contractors, and owners in preparing their positions in these disputes and claims. Clearly, there is a strong return in avoiding claims and disputes in environmental projects.

This chapter looks at the reasons for claims and disputes and discusses innovative ways for owners, architects, engineers, and contractors to avoid or reduce the costs associated with them. Part One presents some recent theories to conceptually understand the origins and factors underlying changes, claims, and disputes. Part Two builds on the theory of Part One and outlines specific activities that can be used to avoid or mitigate changes, claims, and disputes on environmental projects.

PART ONE: CHANGES, CLAIMS, & DISPUTES THEORY

3.2 WHY CHANGES, CLAIMS, AND DISPUTES EXIST

Before entering into a discussion on avoiding changes, claims, and disputes, it is wise to describe what they are and try to understand why they exist. In contract management, *changes* refer to the process of modifying the contract to: (1) add, delete, or revise the scope of work under the contract; and (2) adjust the contract price and contract time, including direction to accelerate the work. *Disputes* generally arise from disagreements about contract interpretations (including the recognition or pricing of changes). *Claims* refer to the formalization of a dispute. Contractor-generated claims describe a situation wherein the contractor gives written notice that a condition exists that requires an adjustment in the contract time and/or price and the owner fails to recognize it by initiating a change order or modification. Some claims are owner-generated in response to some perceived failure of the contractor to perform the work as specified in the contract.

Construction contract literature provides several lists of common changes, disputes, and claims.

Clough (1960, 1969, 1979) lists eight different types of changes: additions to or deletions from the contract; modifications of the work; changes in the methods or manner of work performance; changes in owner-provided materials or facilities, changes in contract time requirements; corrections in the drawings and specifications; change in owner requirements; and changed conditions (also known as *differing site conditions*).

Civitello (1985) cites 10 reasons for change orders: design errors; changes in market conditions; change in the owner's requirements; uncovering of undisclosed existing conditions; uncovering of unknown existing (latent) conditions; suggestions to initiate better, faster, or more economical construction; change in designer preference; discrepancies in contract documents that describe situations contradicting the intent of the project; change in external requirements; and final coordination with not-in-contract (NIC) equipment.

Driscoll (1971) cites eight types of contractor claims: scope changes; constructive change orders; errors and omissions; accelerating and expediting; suspension of work and work stoppage; access or availability of site; interferences, disruptions, and delays by other contractors (at the site); and delays caused by strikes and acts of God. Driscoll also identifies five owner-based claims: failure of contractor to complete on time; liquidated damages and penalties; materials not to specification; defective work and damages; and property damages.

Clough asserts that claims and disputes arise from a variety of origins: interpretation of the Contract; what constitutes extra work to the contract, payment for changes, extensions of time; and damages for owner-directed acceleration, costs occasioned by owner-caused delay, defective drawings or specifications, changed conditions, etc.

In environmental projects all these changes, claims, and disputes can occur. However, the most common appear to be related to design errors (errors, omissions, and discrepancies) and differing site conditions. In reviewing trends in public capital programs, Ponce de Leon (1990) asserts that the breakdown of claimed sums is generally as shown in Figure 3.1; this pie chart shows that schedule extensions and productivity impact claims amount to more than 50% of the claimed sums.

3.3 CHANGES, CLAIMS, AND DISPUTES (CCD) MODEL

The previous lists seem to indicate that claims, changes, and disputes are derived from a complex and sometimes confusing contract environment. However, with some global or macro-oriented thinking, a simple model can be used to understand why claims, changes, and disputes exist. This model, graphically shown in Figure 3.2, is called the Changes, Claims, and Disputes (CCD) Model.

This simple, conceptual model looks at claims, changes, and disputes as a result or *effect* of a multidimensional process. This process can be further subdivided into *generic causes* and *intensity factors*. Generic causes,

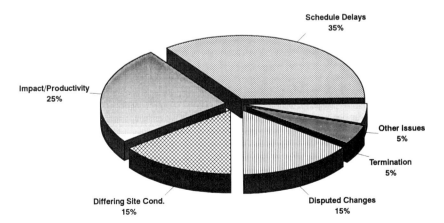

Figure 3.1   Profile of claims on public projects; percentage of claimed sums.

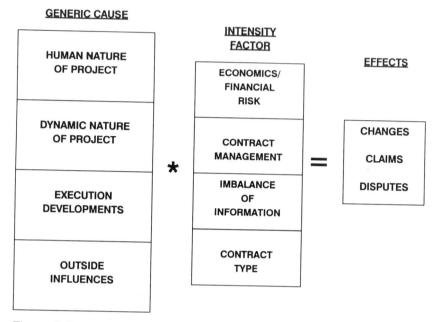

Figure 3.2   Changes, claims, and disputes model.

just as the term infers, are attempts to classify, in a general sense, the main causes for changes, claims, and disputes on any construction project. Intensity factors are factors in the project management process that can either mitigate or accentuate the severity of a change, claim, or dispute.

In Figure 3.2, generic causes and intensity factors are shown as elements of a theoretical equation leading to changes, claims, and disputes. A discussion of each of the generic causes and intensity factors follows.

3.3.1 Generic Causes

The model asserts that there are four separate types of generic causes: (1) the human nature of the project, (2) the dynamic nature of the project, (3) execution developments, and (4) outside influences.

The Human Nature of the Project

Although projects tend to use the latest technologies in design, construction, and controls, the daily decisions and communications still depend on people. Building on the adage, "To err is human," many of the occurrences of changes, claims, and disputes can be explained by the human nature of the project.

Types of changes related to the human nature of the project include:

- Design errors and ambiguities. These would include contradictions, discrepancies, inconsistencies, and flaws in design documents.
- Discrepancies in the contract documents that describe situations contradicting the intent of the project.

The obvious cure for this generic cause is to use a stronger, more timely, quality control scheme to weed out more of the errors and ambiguities. As discussed later, this is the objective of *claims avoidance reviews* of design, bid, and contract documents.

The Dynamic Nature of the Project

Environmental projects, like all projects, are dynamic: New technologies evolve. The owner wants the best facilities for the funds available. Market conditions play havoc with availability of specified equipment. The second generic cause, termed the *dynamic nature of the project*, explains the following changes, claims, and disputes:

- Changes in owner's requirements cause addition or deletion of scope.
- New technologies or market conditions cause change to intended facilities: materials unavailable, new alternatives to specified materials, or new products available.

The best cures for this generic cause are anticipation and quick response. Specifically, solicit owner involvement early in the review process and react to dynamic developments sooner rather than later as the project unfolds. For example, a request to add a second train of clarifiers in a water treatment facility has its minimum effect at the early stages of design rather than at the bidding or construction phase of the project.

Execution Developments

Some changes, claims, and disputes arise from events or decisions experienced during execution of the project versus a difference in the specified facilities as a result of human error, extended scope, or technological developments.

Examples of changes, claims, and disputes resulting from execution developments include:

- All types of differing site conditions (also known as *Changed Conditions*) whether these conditions were known or unknown at the time of contract award.

- More economical means, methods, techniques, or sequences of construction.
- Problems inherent in *fast tracking* or overlapping design and construction activities.

There are selective cures to eliminate claims resulting from execution developments. For example, additional investigations (such as soil borings, material testing, etc.) can be used to clear uncertainties regarding existing site conditions. More frequent quality reviews may be necessary for projects that are on a fast track. Constructability reviews with the contractor at the early stages of the contract can identify more economical construction techniques before they become an issue.

Outside Influences

The previous three generic causes discussed interactions between the three classic participants in the project (i.e., owner, designer, and contractor). The fourth generic cause, termed *outside influences*, is concerned with events or decisions by outside parties, which affect the project.

Examples of claims, changes, and disputes from Outside Influences include:

- Effects of legal or statutory changes on the project (revisions to applicable codes, new clean air/water acts).
- Instances of force majeure: acts of God, floods, abnormal weather, wars, strikes, etc.
- Delays to materials and equipment not procured by the contractor.

Similar to the dynamic nature of the project, the best cures for outside influence are anticipation and early response. Frequent expediting and visits to vendor shops by owner and designer personnel often prevent unexpected slippages in delivery dates for materials and equipment. If new legislation is expected, contingent pricing may be used to cater for its effects.

3.3.2 Intensity Factors

In comparing projects, why is it that similar changes can have markedly different effects? Why do some events or changes escalate into a dispute or a claim while others do not? To help explain these questions, the CCD Model contends that there are intensity factors that positively or negatively influence the impact of an event generated by one (or more) of the generic causes. Currently, the model identifies four intensity factors: economics/financial risk, contract management, imbalance of information, and contract type. Each factor is discussed in further detail.

Economics/Financial Risk

When changes, claims, and disputes appear, inevitably the cost aspect becomes paramount. The economics/financial risk intensity factor recognizes the cost and schedule implications of changes, claims, and disputes. This factor considers economic and financial status from a number of viewpoints:

- Participant financial position: Of the owner, designer, and contractor, who is financially strong or weak?
- Project financial position: What is the status of the project from each participant's view? Who is operating in a *loss* position versus a *gain* position? Who "left money on the table" at contract award and how much was left in relation to changes, claims, or disputes?
- Geographic macroeconomic position: Is the project being executed in a recessionary time or a period of growth? What is the outlook for future projects for each of the participants?
- Cost magnitude: How does the value of the change, claim, or dispute relate to the value of the contract or value of current control budgets?
- Schedule implications: What is the financial risk or reward to each of the participants in completing the contract on time?

The underlying premise for the economics/financial risk factor is that the larger the financial loss or risk, the higher the probability that the losing party will aggressively resolve the contract issue (such as a change, claim, or dispute) in his favor.

Contract Management

The second intensity factor, termed *contract management*, acknowledges the quality of the relationships and interactions among the parties in the contract during the life of the project. Key considerations include:

- Amount of adversarial versus teamwork approach in day-to-day activities
- Ability of all parties to communicate and to listen
- Strength and understanding of the contract language
- Quality of planning and scheduling
- Equitable treatment of previous contract issues
- Quality of change/claim administration

The underlying premise in the contract management factor is that the better all parties work together, the higher the probability that a contract issue (such as a change, claim, or dispute) will be resolved quickly and efficiently.

Imbalance of Information

The third intensity factor, called the *imbalance of information*, recognizes that information is crucial for decision making. With an imbalance of information, it is difficult making equitable and fair decisions. In a claims or changes situation the types of information that should be in balance would include: change order pricing or actual cost data, schedule analysis of delay impacts, facts to demonstrate entitlement of the change, interpretation of contract language, etc. Often one party has more information regarding the issue than the other (or at least believes so). When this imbalance, whether real or perceived, occurs, there exists a significant obstacle in effective resolution of the change or claim.

From a conceptual standpoint, the underlying premise in the imbalance of information factor is that the closer the parties in the contract are to a balance of information the higher the probability that a contract issue will be resolved quickly and efficiently.

Contract Type

The fourth intensity factor relates the type of contract (reimbursable cost, unit price, or lump sum) to the likelihood of claims and disputes. Each contract type carries a different level of risk. Hence, it is understandable that reimbursable cost contracts will typically generate few claims, unit price contracts may yield claims when actual quantities deviate significantly from the bidding basis, and lump sum contracts may generate a substantial number of contract issues, claims, and disputes depending on a variety of factors such as the risk-sharing posture adopted.

The contract type intensity factor is not intended to justify exclusive use of reimbursable cost contracts but rather to recognize that, given the same contract issue (for example, a differing site condition because of unexpected boulders in an area to be excavated), the resolution of that issue may be easier on a project where a reimbursable cost contract is in place rather than a unit price or lump sum contract.

3.4 USING THE CCD MODEL TO IDENTIFY CLAIMS AVOIDANCE ACTIVITIES

The Changes, Claims, and Disputes (CCD) Model asserts that a claim, change, or dispute is the result of one of the generic causes and its severity is determined by the interaction of the intensity factors. In terms of effort versus payout for a claims and changes avoidance thrust, the priority follows the CCD Model: (1) to eliminate or reduce the amount of claims or changes, look at attacking the generic causes of changes; (2) to mitigate changes or claims, attempt to obtain the positive aspects of the intensity factors; and (3) understand that, given the execution basis for a given project, some elements of the CCD Model cannot be influenced.

3.4.1 Attacking the Generic Causes

Of the four generic causes, the human nature of the project can be the most influenced by an innovative claims avoidance program. Through specific quality assurance activities, more human errors can be eliminated. Reviews for constructibility, bidability, claims avoidance, claims exposure, and design decisions accomplish the same objective: to improve the design intent and contract documents for construction by flushing out errors. This is where there can be the highest return.

As for the other three generic causes, changes arising from them tend to be more difficult to eliminate. Rather, emphasis should be placed on recognition, anticipation, and early implementation (which are attributes of good contract management—an intensity factor in this model).

In some projects, the effects of the dynamic nature of the project can be minimized by strong project control leadership by owner personnel.

3.4.2 Influencing the Intensity Factors

In contrast to attacking the generic causes, each intensity factor can be positively influenced to help to mitigate the impacts of changes, claims, and disputes:

- In understanding the effects of the economic/financial risk, owner or engineer estimates can identify areas of financial risk in contractor bids. Reviews of financial status of prospective bidders can eliminate possible insolvent bidders.
- The contract management factor can be improved through team building, "risk sharing" contract language, and recognition, anticipation, and early response to situations that lead to claims and disputes.
- The imbalance of information factor can be improved by good organization of project documents by all parties, regular analysis of payment requests and schedule submittals, and maintenance of appropriate daily records and logs.

- Through early recognition of possible changes, contingent prices or requests for bid prices for alternate designs can be used to improve the contract type factor.

PART TWO: SPECIFIC METHODS TO AVOID OR MITIGATE CHANGES, CLAIMS, AND DISPUTES

As mentioned in the introduction, Part Two builds on the CCD Model discussed in Part One and outlines specific activities, procedures, or reviews that can be used on environmental projects to avoid or mitigate common changes, claims, and disputes. Figure 3.3 provides a graphical outline of the methods described in Part Two as they relate to an overall claims avoidance plan.

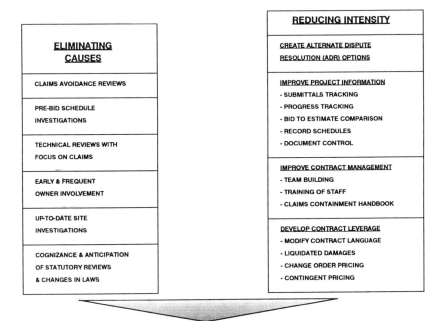

AVOID CHANGES, CLAIMS & DISPUTES

Figure 3.3   Core of a changes and claims avoidance plan.

3.5 CLAIMS AVOIDANCE REVIEWS

Claims avoidance reviews are systematic reviews of design and contract documents that focus on identifying areas susceptible to changes, claims, and disputes. These reviews assist in a total quality management (TQM) program to reduce human errors in these documents. Claims avoidance reviews critique the constructibility and biddability of these documents.

For a standard or typical environmental project, there should be at least three separate claims avoidance reviews:

- Avoidance review of design documents
- Avoidance review of bidding documents
- Post-award review with successful bidder

Figure 3.4 illustrates the timing of these reviews against a typical schedule for an environmental project. The optimal timing for the claims avoidance reviews for design and bidding documents is when there is sufficient information available to perform an effective review plus enough time to recycle designs and bid documents without causing a delay to the master

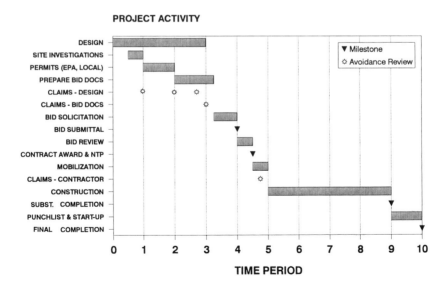

Figure 3.4 Typical schedule for environmental projects showing timing of claims avoidance reviews.

schedule. Ideally, claims avoidance reviews of design documents should occur at the 70% and 95% complete milestones and the reviews of bid documents should occur several weeks before solicitation of bids to cater for this requirement.

The post-award review with the successful bidder (i.e., the contractor) should occur soon after contract award. This timing takes advantage of the so-called honeymoon period that occurs just after contract award and before substantial work is performed at the project site.

3.5.1 Avoidance Review of Design Documents

Claim avoidance reviews of design documents look at the content as well as the implied instructions given to the contractor. Some examples of review emphasis are:

- Investigate the availability of specified equipment items or key materials as well as an outlook for obsolescence. For instance, in water treatment plants, identify pumps and drivers associated with unique capacity and/or head requirements, those containing exotic materials (any materials other than cast iron or carbon steel), or those of unique design. Check the number of vendors who meet these requirements with machinery specialists on the design team. Review the necessary delivery/lead times.
- Compare the present plans, specifications, and designs to the track record of similar designs in the recent past for claims, changes, and disputes to determine whether history will repeat itself.
- Assess the possibility that a competent contractor can meet the specified performance requirements, sequences of work, etc. For example, investigate the ability of local contractors to meet the compaction requirements for hazardous waste landfill liner materials (from a material as well as a construction equipment availabilty). Also, often tie-ins of new to existing water treatment facilities require coordination with low flow operations, so inquire as to how this is outlined in the design.
- Review the relevance of site investigations data to the design intent. Consider whether the site investigations data has become obsolete because of time delays or disturbances caused by other projects at the site. For instance, geographically relate the location and frequency of soil borings to the location of new facilities or underground work (such as tunnels, culverts, conduits, etc.) Also, for work in existing buildings, inquire whether tests for ACMs (asbestos-containing materials) have been performed and what procedures are in place for removing ACMs.

- Compare cross references in plans and specifications, and between plans and specifications to identify errors. Trace references to code requirements to determine whether the references are current.

3.5.2 Avoidance Review of Bidding Documents

Claim avoidance reviews of bid documents primarily look at the contract language and special instructions of the bid documents. When a claims avoidance review of the design documents has not occurred, this review could also cover the design documents (using the procedure for Review of Design Documents). Some areas of review emphasis are:

- Inclusion of appropriate disclaimers regarding subsurface conditions indicated by soils reports to avoid unreasonable reliance on those reports by bidders.
- Consider provisions stipulating that float in the contract schedule can be used as a bar to time extensions.
- Review the contract treatment for variations in estimated quantities. For example, investigate how the contract would handle an additional 3 feet of depth in excavation.
- Include in the dispute clause, a provision that requires formal submittal of a claim as a prerequisite step to filing a lawsuit. Also, in the event of litigation, venue should be restricted to local courts.
- Review material or equipment substitution provisions in order to disallow equipment items that do not improve the life cycle cost of as-bid equipment and to ensure that substitutions consider all secondary effects such as delays.
- Incorporate bid dispute provisions aimed at clarifying procedures related to bid withdrawal, objections to award of a contract, and conditions that may warrant disqualification of a bidder.
- Review the clarity in language used to determine the order of precedence in the contract documents when there are conflicts or inconsistencies.
- Evaluate the need for a CPM-based progress schedule specification.
- Review provisions to handle defective work, particularly when a variance is acceptable but the language is not clear about the recovery for delay.
- For environmental projects involving the handling, storage, or removal of hazardous, toxic, or asbestos-containing materials, add language covering prequalification of the contractor or subcontractors with appropriate governmental agencies as a requirement of the bid.

3.5.3 Post-Award Review with Successful Bidder

A substantial number of potential claims can be avoided through quality control and constructibility reviews of the contract documents during the *contract start-up* phase by key owner, designer, and contractor personnel (including appropriate subcontractors and suppliers). During these reviews the majority of any significant errors, omissions, ambiguities, or inconsistencies in the plans and specifications should become apparent. Reviews should culminate with lists of (1) issues requiring interpretation or clarification by the owner and designer, (2) specific changes needed to overcome problems discovered, and (3) possible changes that would be desirable depending on the impact on contract price and time.

To execute these reviews in an optimal way and to avoid an inordinate load on staff, these reviews can be approached by a trade or discipline fashion. Meetings to share information can be planned on a sequential basis starting with site preparation and earthwork and continuing through concrete, tunneling, steel work and superstructures, piping and mechanical, and so on.

3.5.4 Staffing Requirements for Claims Avoidance Reviews

Each claim avoidance review requires: (1) a technical person or persons who are knowledgeable in the type of work described by the contract, and (2) a project management specialist who understands contract administration and the effects of changes, claims, and disputes. In addition, where practical, it is often useful to have a key owner's staff member available to understand and relate the project's coordination issues with the owner. The number of individuals involved is dependent on the size of the project, the complexity of the technology used in the project, and the amount of time scheduled to perform the review.

3.5.5 Considerations for ''Aged'' or ''Rushed'' Projects

Aged projects are those in which, for one reason or another, the normal progression of project development (shown in Figure 3.4), has been postponed. A frequent example of an aged environmental project occurs when the design and bid documents become dormant as the project awaits EPA or other governmental funding or approval. In contrast to aged projects, rushed projects are those in which the progression of project development is accelerated. Examples of rushed projects include those that are on a "fast track" (i.e., there is an overlap in design and construction activities), or those being modifiedy by a value engineering concept.

Aged or rushed projects pose a higher susceptibility for changes, claims, and disputes and the following additional review activities can be used to avoid some of the problems inherent in these types of projects.

- Investigate how the current location and scope of facilities relate to assumptions used in the previous design. If project facilities have been relocated, check the relevance of soil borings or other site investigations against the new location. Also, visit the site to ascertain whether any refuse, spills, or other unexpected materials exist at the new site.
- For aged projects, check the validity of as-built conditions to avoid changes caused by deterioration of existing facilities or construction of new facilities adjacent to the project site.
- In rushed designs where upgrading of equipment items occurs, check that recommended sequences of construction, specified temporary scope of work, and existing facility connections have been modified according to the needs of the new design.
- For aged projects, check front-end and regulatory sections of contract documents to identify obsolete insertions. Replace with the latest updates.

3.6 TECHNICAL REVIEWS WITH A CLAIMS EXPOSURE FOCUS

To complement separate claims avoidance reviews and convey a "claims avoidance attitude," normal technical reviews of design and contract documents should also contain a claim exposure focus. Many changes and claims can be eliminated by making the correct decisions and obtaining the owner's preferences at the early stages of the project.

Some suggestions for developing a claim exposure focus in technical reviews include:

- Playing "devil's advocate" and challenging design decisions. This technique is often most successful when adopted by design professionals who are not intimately involved in the daily decision-making process.
- Schedule meetings with owner representatives at key milestones in the design process. Allow operations and maintenance personnel as well as project executives to review the status of the design. Discuss the owner's plan for handover of the facilities. Many owner preference change orders can be eliminated by effective design reviews.

- Compile a "Handbook of Avoidable Changes" from recent projects and compare the proposed design against the handbook.

3.7 PRE-BID SCHEDULE INVESTIGATIONS

Another frequent review process that looks at the constructibility issues of the project is a Pre-Bid Schedule Investigation. This review uses the design or bid documents and determines: (1) a recommended contract period and level of liquidated damages; (2) milestones for contract award dates; (3) recommended or required sequence of work restraints, (4) durations for owner and designer activities during construction; and (5) requirements for a CPM-based scheduling specification that will correlate the general conditions and requirements of the contract.

In addition to these five general goals of a pre-bid schedule investigation, there are a number of other specific areas that should be reviewed in this investigation:

- From a construction logistics standpoint, look at the site conditions. For example, for an expansion of an existing waste water treatment plant, consider accessibility of the site by construction and delivery vehicles, determine the largest loads that can safely be brought in and out of the site, look at the availability of laydown areas and temporary parking, and inquire about restrictions in traffic flow both inside the site as well as in the local community.
- Review the effect of weather conditions on the project. For example, is there a need to be watertight before certain work activities can continue in a building enclosure? What are the regional norms for seasonal disruptions such as hurricanes or harsh winter weather where restrictions on concrete, masonry, or painting activities can occur?
- Review the definition of substantial completion and final completion for the project. Investigate whether there is a "phased" or "staged" handover. Identify any activities that can occur between the contract milestones of substantial and final completion (such as punchlisting, landscaping, final instrumentation checks, etc.).
- When a phased or staged handover is required, review the responsibilities for maintenance of facilities turned over (prior to final completion).
- For projects that will use multiple prime contractors, identify and review the interactions between contracts. Inquire about the specific interface coordination between the contracts. Review the schedule recog-

nition of these interfaces. For example, if one contractor will be constructing an outfall for a water treatment plant while another will be building the expanded facilities, look at the coordination and scheduling requirements for the outfall connection. If a mechanical contractor is placing floor penetrations and an electrical contractor is placing conduit within a concrete pour, look at the scheduling and coordination of these activities.

The previous activities in a claims avoidance plan identified ways to eliminate causes of changes, claims, and disputes. This section discusses activities that can reduce the severity of changes, claims, and disputes on a project. There are four major themes to be discussed: (1) creating Alternate Dispute Resolution (ADR) options, (2) improving the Project Information process, (3) improving the Contract Management process, and (4) developing Contract leverage.

3.8 ALTERNATE DISPUTE RESOLUTION (ADR) OPTIONS

Many owners, designers, and contractors feel that litigation is a costly process that interrupts normal business operations and impacts business relationships, *alternate dispute resolution* has emerged as a popular option to resolve contract issues on construction projects. ADR can be determined at the time of the dispute or outlined as part of the contract. Muller (1990) lists nine different types of ADR methods or techniques: negotiation, mini-trial, contract dispute review boards, rent-a-judge, mediation, court-appointed masters, expert resolution, binding arbitration variations, and nonbinding arbitration variations.

Each type of ADR has it merits and there is likely to be an option that will best suit the parties in dispute. In considering an option to use as part of a claims avoidance plan, the following criteria should be considered: precedence of case law in the locality; availability of staff, experts, or arbiters for mediation teams or dispute boards; recent ADR experiences; and protection of expert work products from discovery and freedom of information requests.

3.9 IMPROVING PROJECT INFORMATION

The CCD Model contends that information is a key factor in mitigating changes, claims, and disputes. There are two main types of information as targets for improvement: (1) basic project information, and (2) changes, claims, and/or dispute entitlement documents. Each target type and mechanisms for improvement are discussed further.

3.9.1 Basic Project Information

- *Bid to estimate comparisons*: Prior to bidding, it is useful to have the owner or designer compile an estimate used to compare contractor's bids. Suggested details for each component or facility include: bulk quantities, direct manhours, labor costs, unit installation rates, equipment costs, and material costs. On developing the indirect estimate consider job site office costs, construction equipment charges, and contractor's gross margin to cover home office overhead and profit.

  As part of the contract bid, require each bidder to provide a bid breakdown that corresponds to the estimate discussed above. In lump sum contracts, require that the bidder provide this breakdown as a prerequisite to evaluation of the bid.

  A comparison of the bid to the estimate will provide cost information to recognize areas of strength and weakness in the bid. This knowledge can be useful in understanding the contractor's financial position and areas of potential exposure on the project.

- *Submittals tracking*: Develop a list of all shop drawings and other contractor submittals to be reviewed and approved during the construction phase. Implement a submittal tracking system that will control the flow of documents from the contractor to the reviewing party. This system will enable effective planning of reviews, potential delay avoidance, and provide a contemporaneous record of submittal events, which can be used in evaluating claims and disputes.

- *Progress schedule tracking*: Schedule control is essential in any construction project to understand what has been done, what remains to be done, and when the remaining work will be done. There are two phases in good schedule control: (1) development of control tools, and (2) periodic monitoring of progress and updates to the schedule.

  Requirements for project schedules are usually specified in the contract. With the advent of powerful personal computers and appropriate software, CPM scheduling is commonly specified along with a host of reports to identify activity relationships, resource requirements, rate of progress, etc. During the first 10% of the construction duration, the contractor is preparing the initial detailed CPM schedule (also known as *detailed work plans*) and creating the specified reports. It is during this development that conducting working-level reviews of the contractor's schedule is crucial in understanding the details of the contractor's plan to execute the work. Key areas of review would include: resource leveling, adherence to specified sequences and constraints, consideration of owner's activities (including submittal reviews), compliance with contract times, manipulation of float, and composition of periodic reports.

Upon successful establishment of a control schedule, the emphasis centers on periodic updating of the schedule and monitoring schedule performance. Similar to the reviews for the initial progress schedule, the updated schedule should be reviewed for: changes in sequencing of construction activities, incorporation of all contract time modifications that are the result of approved contract changes; verification of actual dates for completed or in progress activities; revisions to curves to report rate of progress; and compliance of forecasted completion dates against the contract times.

3.9.2 Changes, Claims, and/or Dispute Entitlement Documents

- *Document control*: Environmental projects, like any other construction project, generate a plethora of documents. Some pre-investment in developing a computerized document control system may pay dividends when changes, claims, and disputes arise. Through appropriate indexing of documents, a quick recall of facts that are pertinent to evaluating changes, claims, and disputes can be accomplished.
- *Video surveys*: Conduct a video survey of the site during the period between bid advertisement and bid opening. In addition, when there is a point of contention, video the actual conditions at the site in question. These surveys can be used as demonstrative exhibits during the resolution of the dispute.
- *Record or "as-built" schedules*: Legal precedence shows that delay disputes can be resolved by analyzing critical paths in the as-planned versus the as-built schedule. Why wait until litigation? A proactive claims avoidance plan should include the development of an independent record schedule to show the actual start and finish dates of each activity, the actual logic sequencing, disruptions in the planned work, and extra work related to change order negotiations. An independent record schedule can be used to compare the contractor's submittal of a record schedule as well as the evaluation of a request for a time extension.

Important features of an effective record or "as-built" schedule include: clear distinctions between original, modified, and disputed work; as-built activities in the same level of detail as the contractor's schedule; identification of idle time, crew shifts, and remedial work; sequence of activities in negotiation of change orders; and correlation of data from project logs for submittal approvals and material deliveries to site or storage. In the case of multiple contracts, the interface between prime contracts (such as site access, approval of shop drawings and other submittals, and material deliveries) should be identified.

3.10 IMPROVING THE CONTRACT MANAGEMENT PROCESS

Another important aspect of claims avoidance is the approach used to manage the contract. It is the contention of the CCD Model that claims and disputes can be mitigated through teamwork, a claims avoidance attitude, and sound project management skills. Using this contention, a selection of innovative ways to promote improvements in contract management include: team building with a focus to avoid claims; additional training of staff; and development of a claims containment handbook.

3.10.1 Team Building with a Focus to Avoid Claims

Team building with a focus to avoid claims has two thrusts: (1) to break down communication barriers and adversarial relationships between owners, designers, and contractors, and (2) to create an attitude that eliminates avoidable costs associated with changes, claims, and disputes. Specific activities that can be helpful include:

- *Mission Statement*: Development of the project's goals and objectives in regard to avoiding claims and disputes can be stated in a mission statement that can be disseminated among the project team. An example of a typical mission statement, called the *ten canons of claims and disputes avoidance* follows:

  1. Minimize/eliminate delays of early project activities.
  2. Reduce the cost and schedule impact of changes.
  3. Create a risk-sharing posture with contractors.
  4. Reduce ambiguities in contract documents.
  5. Manage the contract with focus on claims avoidance.
  6. Be in an equal position to negotiate costs of changes and disputes.

7. Be in an equal position to negotiate schedule effects of changes and disputes.
8. Provide an effective ADR (alternative dispute resolution) procedure or process.
9. Enhance your chances of success in litigation (if necessary).
10. Periodically self-audit the effectiveness of the claims avoidance program.

- *Periodic changes, claims, and disputes meetings*: For large contracts (greater than $50 million), periodic meetings between key owner, designer, and contractor personnel can be productive in resolving issues before they escalate into disputes. Discussion can center on: investigating more efficient ways to incorporate changes into the construction sequence, outlining specific actions needed to negotiate cost and schedule effects of a change or claim, or defining documents necessary for entitlement of a change or claim.

3.10.2 Training of Staff

Consistent with the goals and objectives of a claims avoidance program, additional training of the project staff is often needed to upgrade key project personnel and senior management. The goal of the training is to develop skills required to avoid, mitigate, and resolve contract issues. Specific training seminars would address:

- Project goals and objectives in regards to the claims avoidance program. A mission statement such as the earlier example can serve as the focus of the seminar.
- Knowledge of the contract documents and language used. By identifying key contract clauses, equitable and consistent interpretations of the contract documents can be used to contain many claims and disputes. Examples of areas to review would include: contract change pricing schemes, risk-sharing provisions, differing site conditions, provisions for time extensions, alternate dispute resolution procedures, etc. A review of current legal precedence underlying each topic would also be useful and enlightening.
- Knowledge of unusual features of the project design. Examples of topics to cover would include: subsurface and other site investigations; definition of substantial and final completion including the contract time milestones; specified construction sequences, means, or methods; re-

quirements for staged or phased handover; and owner/designer activities during construction.

- Contract management procedures including understanding of specified responsibilities. Topics to be reviewed would include: daily logs, progress schedule analysis, change order administration, project correspondence, shop drawing and other submittal approvals, payment certificate analysis, etc.
- Procedures for handling requests for information (RFIs) and how to deal with common disputes arising from questions, format, and decisions in these regards.
- Management of a change, claim, or dispute. Topics to be reviewed would include relevant contract language, flow chart of responsibilities to assess and resolve the issue, schedule analysis required to determine time extensions, change order estimating, etc.
- Management of alleged differing site conditions, taking into consideration the contract provisions, site investigations (such as soils reports), and handling of hazardous, toxic, or asbestos-containing materials.
- Improving negotiation fundamentals and communication skills including role-playing examples of mock disputes.
- The value and need for team-building and professional behavior with all project personnel.

The extent of claims avoidance training for a project team member should be related to the influence that the individual has to avoid changes, claims, and disputes. However, it is important that all members of the project team be familiar with: (1) the intended design concept for the project, (2) the contract documents, (3) goals and objectives of the claims avoidance program, (4) the project schedule, and (5) norms of good practice regarding claims containment.

3.10.3 Claims Containment Handbook

For very large projects (greater than $100 million), a "Claims Containment Handbook" can serve as a supplement or byproduct of the formal staff training on claims avoidance. The handbook would be a reference book for all project personnel who deal with changes, claims, or disputes. The handbook would address the mitigation, processing, analysis, and settlement of disputes and claims.

Similar to the training seminars, a Claims Containment Handbook would include:

- An explanation of claims containment contract provisions/clauses and their risk-sharing premise. Specific provisions in the project's contract should be copied and discussed.
- Procedure for administration of contractor schedule submittals. Discussion of issues such as review/approval of schedule, noncompliance, and float suppression should be included.
- Procedure for preparing independent record or as-built schedules.
- Procedure for the administration of change orders, claims, and disputes. This procedure should include: flow charts and responsibility matrices for the assessment and resolution of contract issues, guidelines for pricing changes as stated in the contract, guidelines to determine time extensions, etc.
- Procedures for handling requests for information (RFIs) and how to deal with common disputes arising from decisions in these regards.
- Key regulatory and project documents.
- Glossary of project terms and selected documents (such as recent legal briefs).

3.11 DEVELOP CONTRACT LEVERAGE

The contract is the most important factor in the claims avoidance program. The contract describes the intent of the project, time deadlines for its completion, the price that should be paid, etc. Within this chapter, a key concept has been implied regarding the contract: contract provisions can be adopted, modified, or enhanced to help avoid extraneous costs caused by changes, claims, and disputes. This section discusses innovative ways to "leverage" the contract to avoid these extra costs.

3.11.1 Key Modifications to Contract Language

Throughout this chapter, numerous references have been made to include specific clauses, provisions, or specifications in the contract to assist in the claims avoidance or containment process. A brief checklist of these references follows:

- Bid dispute provisions: procedures for bid withdrawal, objections to award of a contract, and conditions that may warrant disqualification of a bidder.

- CPM progress schedule specification: comprehensive progress scheduling provisions that address initial progress schedule submittal, frequency of periodic submittals (i.e., updates), content of submittals, short-term lookahead schedules, compliance with contract times, and float suppression.
- Comprehensive change order pricing provisions.
- Audit clause to allow access by owner representatives to contractor's monthly job cost reports to verify costs associated with claim and dispute estimates.
- Dispute provisions: outlining alternate dispute resolution (ADR) procedures, venue for litigation (normally local courts), requirement that formal submittal of claim is a prerequisite step to filing a lawsuit.
- Material or equipment substitution provisions: disallow equipment that does not improve life cycle costs and ensure that substitutions consider all secondary effects such as delays.
- Quantity variation provisions for estimated quantities in unit price contracts.
- Additional bid information: requirements for a breakdown of lump sum bid price into specific components; submittal of preliminary schedule after bid opening but before contract award.
- Owner review of contractor submittals: specify duration for owner's review and timing of subsequent reviews if submittal has to be revised.
- Change order notice procedure: contractor required to number requests for change orders consecutively.
- Daily field log: specify requirements for contractor's daily field log.
- Waiver provisions: address reservation of rights, waiver of future claims regarding a change order, etc.
- Backcharges provisions for unanticipated owner's costs such as overtime inspection.
- Order of precedence in contract documents when there are conflicts and inconsistencies.
- Float as a bar to time extension clauses.
- Use of soils reports: appropriate disclaimers to avoid unreasonable reliance on these reports by bidders.
- Termination provisions to allow termination of contract when delays jeopardize the viability of the project schedule.

In addition to these modifications to the contract, some consideration should be given to using supplements to standard provisions or completely rewriting them. The decision is likely to hinge on the frequency of use of the documents versus the cost of rewriting. In any event, it is also advised to reduce cumbersome legalese in contract documents.

3.11.2 Liquidated Damages

One common form of contracting leveraging is liquidated damages. Liquidated damages, or "liquidation of damages clauses" as they are known in contract law, are contractual provisions under which the parties agree that unexcused delays will cost the contractor specified sums of money. Liquidated damages clauses must pass two tests for legitimacy: (1) the damages must be *difficult to ascertain* at the time of the contract, and (2) the specified amount must be based on a *genuine pre-estimate* of the probable damages the owner would incur in the event of a delay.

Liquidated damages clauses usually specify an amount of damages per calendar day that the contractor must pay when the contract is not completed by the specified contract time(s). Typically, in recent water treatment projects, there are two amounts specified in liquidated damages clauses: one amount for breaches to substantial completion milestones and the second amount for breaches to final completion. In many cases, the liquidated damages for substantial completion is much higher than final completion because more damages occur in postponing substantial completion (because the owner can often use the plant at substantial completion).

The liquidated damages clause is an exclusive remedy for late completion; other clauses in the contract allow the owner to recover damages for contract breaches such as defective work and abandonment.

Although the concept of liquidated damages is not a recent one, there are a number of innovative approaches that make the clauses more enforceable as well as equitable:

- Use probalistic estimating techniques to assist in quantifying the amount of liquidated damages. This is a three-step process: (1) identify the owner's possible damages, (2) quantify the damages based on a range of delays, and (3) assess probabilities for the range of delays.

  Items to be considered in identifying owner's possible damages include: extended staff costs of owner's representatives assigned to the project, excess financing costs, use of temporary facilities and operations, continuation of facilities and operations to be replaced by the new project, delay damages to follow-on or succeeding contracts, and fines levied by the EPA or other regulatory agents if the project is delayed.

For each selected range of delays (e.g., 0 to 20 days' delay, 20 to 40 days' delay, etc.), each identified damage is estimated. Some damages, such as extended costs of owner's representatives or use of temporary facilities, will vary directly with the amount of delay. Other costs, such as follow-on or succeeding contracts and regulatory fines, depend on their time relationship with the project. For example, if there is 50 days' float between succeeding contracts, there will be no damages or disruption costs of the secondary contracts until the project has been delayed by more than 50 days.

Finally, a probability can be assessed for each range of delay selected. Ideally, this probability should be related to an empirical analysis of previous projects. For example, a statistical analysis of recent projects for similar hazardous waste cleanup projects or waste water treatment plant expansions will yield information on the schedule performance of contractors. However, in reality, these probability assessments may become more judgmental due to lack of data or time to carry out a statistical analysis.

- Use appropriate contract language when specifying liquidated damages. The fact that the parties used the words *liquidated damages* in their agreement does not prevent a court from construing the agreement to be one for a penalty. Hence, avoid the words *penalty* or *forfeit* in these clauses.

- In government projects where liquidated damages schedules are available from procurement regulations, use the schedule or an independent assessment but not both. Often courts will rule that the addition of the estimated damages to such a schedule effectively doubles the estimated loss.

- Be specific in choosing the milestones and the liquidated damages assessed to the milestones. Most contracts use the substantial completion milestone as the key date to trigger liquidated damages because the owner generally takes beneficial use of the project at that time and damages assessed thereafter begin to appear to be a penalty rather than compensation for losses.

    Also, if more than one milestone is being used in liquidated damages clauses, use clear language to specify the effect of each liquidated damage upon breach of each milestone. For example, if the liquidated damages are to be exclusive and additive, specify this accordingly.

- Some recent projects, such as the new water treatment and pumping station in Lake County, Illinois, are adding bonus/penalty provisions in addition to liquidated damages to achieve appropriate incentives for schedule performance.

3.11.3 Change Order Pricing

Many projects' contract documents do not provide specific guidelines for pricing changes. Consequently, the parties are left to negotiate the best deal they can get, using previous experience as a guide. This negotiation process can be bypassed by specifying a comprehensive method for pricing changes as part of the contract or agreeing on a method at the onset of construction (i.e., during the honeymoon phase between contract award and start of substantial work at the site).

Specific pricing data would include:

- *Labor costs*: Prevailing wage rates, labor burden factors (i.e., payroll taxes, fringes, etc.), typical craft mixes per discipline, and manhour unit rates can be derived from standard estimating guides such as *Means and Richardson's*. Adjustments may be needed when the contractor's cost basis is not the same as assumed in the standard estimating guide (e.g., foremen as direct labor or field staff).
- *Material/equipment costs*: Guidelines for treating transportation and storage costs, trade discounts/rebates/refunds, and the number of required supplier quotations should be addressed.
- *Construction equipment*: Discuss appropriate treatment of contractor-owned vs. rented equipment, use of estimating standards such as Blue Book, correct pricing of rental equipment (i.e., monthly hire rate when equipment is on site for a month, weekly hire rate when equipment is on site for less than a month but greater than 5 days, etc.).
- *Small tools*: Specify minimum purchase cost of construction equipment versus a small tool (normally $500). Specify pricing scheme for small tools: as a percentage or on a reimbursable basis.
- *Site office and general conditions costs*: In case of compensable time extensions, specify treatment of nonvariable costs (i.e., fixed costs such as building purchases) versus variable costs: only variable costs should be compensable.

3.11.4 Contingent Unit Prices

If the contract documents do not provide unit prices or if alternate prices are required because of perceived changes in the design intent, the contractor can be requested to provide selected unit prices for the uncertain scope of work. For instance, on an expansion to a water treatment plant, unit prices could be negotiated or requested for a variety of activities:

- Additional cubic yards of overexcavation when unsuitable material is encountered at below-grade elevations.
- Supplemental cubic yard rates for disposal of different types of contaminated excavated materials (i.e., ignitable, corrosive, reactive, or toxic).
- A three-foot basis for longer or shorter lengths of structural piling.
- A per-linear-foot basis for leaving sheet piling in place.
- A linear-foot-basis for removing electrical wiring and installing and testing replacement wiring.

3.12 CONCLUSION

Most major projects today are developing some form of claims avoidance plan. The theory and specific activities outlined in this chapter can be used to enhance and develop these plans further. The next logical step in claims avoidance is to develop yardsticks to judge the effectiveness of a comprehensive plan. There are several parameters by which performance can be measured:

- Number of RFIs or clarifications to the design documents. If the claims avoidance program is working, the quantity of RFIs and clarifications should be significantly reduced.
- Changes, claims, and disputes as a percent of contract value. In the introduction, a rate of 15% to 25% was cited as an observed range. A project with a successful claims avoidance program should experience a much lower percentage.
- Estimated savings resulting from a reduction of changes, claims, and disputes versus the investment in staff, training, and other activities in the claim avoidance program.

REFERENCES

Civitello, Andrew M. (1987). *Contractor's Guide to Change Orders*, Prentice-Hall, Inc., New York.

Clough, Richard H. (1960, 1969, 1979). *Construction Contracting.* John Wiley & Sons, New York.

Driscoll, Thomas J. (1971). "Claims." In *Contractor's Management Handbook*, edited by James J. O'Brien, P. E. and Robert G. Zilly, P. E., McGraw-Hill, New York.

Grogan, Tim and Steven W. Setzer (1991). Good News, with Reservations. *Engineering News Record*, August 19: 8–9.

Henderson, CCE, Thomas R., (1991). Analysis of Construction Disputes and Strategies for Cost Minimization. *Cost Engineering*, November: 17–21.

Muller, Frank (1990). "Don't Litigate. Negotiate! *Civil Engineering*, December: 66–68.

Ponce de Leon, P. E., Dr. Gui (1990). "Claims Management for Major Construction Projects." Presentation to Detroit Chapter of NSPE on May 23, 1990.

4
Assessment and Remedial Cost/ Schedule Baseline Development

Marc A. Zocher
Project Time, & Cost, Inc., Albuquerque, New Mexico

Gary E. Thompson
Los Alamos National Laboratory, Los Alamos, New Mexico

4.1 INTRODUCTION TO ENVIRONMENTAL BASELINES

Mankind has realized that in recent history adverse stress has been placed on the environment. Along with this knowledge, we have realized that some action is required to reverse this trend, to remediate areas that due to this pollution pose a risk both to human health and to that of other life inhabiting these areas. Remediation requires careful planning to ensure that resources are wisely spent, and that the polluted area or areas are returned with clean-up goals met and human health risk reduced. The development of environmental baselines represent this planning process for both the assessment and remediation phases of environmental cleanup.

To describe environmental baselines, an understanding of baselines as a planning tool is fundamental. A baseline has two major components. The first part, or the static baseline, is the planning process that culminates in an agreed-upon scope, schedule, and estimate of a project. The key ingredient in this definition is the statement that the plan is agreed upon between all interested parties of the project. The individuals that are typically involved at this stage include project owner(s), contractor(s), subcontractor(s), and, in the case of an environmental project, state and federal regulators. Each baseline contains a plan of the technical work to be accomplished (e.g., to stop a leaking tank and clean surrounding soils), the schedule of accomplishment, and an estimate of the dollars and resources required to complete the work. Similar to construction, environmental remediation projects have all the traditional project components that are subject to project management. However, the three major points of departure from traditional construction baseline development are:

- By their nature, environmental projects contain many more unknowns.
- Regulations are the driver—placing emphasis on schedule compliance rather than cost.
- Traditional estimating support such as cost data bases and cost guides do not currently exist.

All these points must be considered when developing an environmental baseline.

The static baseline is often referred to as a planning "snapshot" at the time of initial agreement on the components of the project. The second aspect of the baseline, often called the *dynamic baseline*, is defined as the point in time when the static baseline is set in motion, and ends when the project is complete. The entire baseline never changes through time; it can only be updated in discrete steps when replanning is required. In other words, the initial "snapshot" plan of the project (static baseline) can be updated by taking another snapshot of the updated plan (as part of the dynamic baseline) as many times as required in the life of the project. These updates are commonly referred to in project management as *change control*, or *rebaselining*. This usually occurs when more is learned about a site, and major revisions to the plan must occur. The entire baseline is shown in equation form below.

$$B_{static} + (B^1_{dynamic} + B^2_{dynamic} + B^3_{dynamic} + \cdots B^n_{dynamic}) = B_{total}$$

where B = Baseline

and each step, $B^n_{dynamic} = B^{n-1}_{dynamic}$ + changes since last update.

In summary, a baseline is like a camera that takes a snapshot, not like a video camera that takes constant footage as the plan is updated. Moreover, you are allowed, under project management control, to rebaseline a project if significant changes occur to the scope as more is learned. A baseline that plans for remediation prior to assessment will need to contain major assumptions on how cleanup will occur because the contaminants, if any, are not known. Once the assessment phase is almost complete, rebaselining may be needed because the most likely remediation technology may not have been considered in the original plan. In environmental work, estimates are often required at those early stages for planning purposes.

4.2 STATIC BASELINE DEVELOPMENT

In order to develop the initial project baseline, four major areas of a project must be examined and developed. These development areas are:

- Overall program management definition
- Technical scope development
- Schedule development
- Estimate, or cost development

All four are important, and good overall program management is as valuable as the three common elements of scope, schedule, and cost. Program management in the environmental arena requires strong management teams because the scrutiny on these projects is extreme. Because environmental projects are highly regulated, this interest in good baseline management should be neither unexpected nor unreasonable.

4.2.1 Program Management

The most important and fundamental step in the development of environmental baselines is the program management definition. Although this step is considered to be an implicit component of any technical baseline development, it should be explicitly dealt with in the complex environmental arena. The unstable nature of the regulatory environment, the impact of new technologies, and the inherent unknowns of site conditions and waste streams are all valid reasons for doing a great deal of planning at the top level. Projects are defined under the umbrella of a program and are considered to be discrete pieces of a program. In private industry environmental cleanup actions and current waste management are typically combined to form one corporate environmental program that is planned and executed by a dedicated department or team. In government these environmental projects are dealt with by the entities that managed the sites needing remediation. In the United States, for example, the U.S. Department of Energy (DOE) has a large cleanup effort underway to remove hazardous and nuclear waste from weapons production plants and laboratories located across the country.

The three major planning and management steps are identifying the mission, developing the project, and documenting and controlling the baseline once the project is set in motion. By employing this top-down definition approach, the planning that occurs from the bottom-up will be done within a logical framework.

The mission definition is a statement of the end product or end condition to be achieved at the completion of the program or during it. This statement provides the boundary conditions that govern further project development and change evaluation. Typically, the mission statement provides a measurable objective for all parties involved in the mission and should be strong, clear, and concise.

An example of a poor mission statement for a corporate waste management program would be "to evaluate the effectiveness of our waste minimization program." How will you know when this is complete (i.e., that the mission has been accomplished)? This subjective approach is better stated as "We will reduce corporate waste streams of hazardous waste by 8% per year and of sanitary waste by 5% per year for the next five years." This mission statement may be a subset of a larger mission for the installation or corporation. The overall corporate waste reduction mission may be 12% per year, but the reasonable mission for this plant is only 8% per year because of the nature of the production there. An environmental restoration project should have an end mission to clean up the site to an agreed-upon compliance level based on land use plans for the area being remediated. This is a good example of the complexity of environmental restoration; at some locations the land use levels have not yet been promulgated. Estimates, schedules, and technical scope rely heavily on the mission and, as this case illustrates, an unsettled mission will be a factor in the confidence of the overall baseline.

The next step in environmental project management involves the key activities of development. These include planning, programming, defining the static baseline, and identifying needed resources. Planning begins with mission definition and continues to greater levels of detail until all required work elements are identified. This process produces program narrative and is developed by the program team and other advocates of the mission. In the case of the government a major advocate is the general public. By obtaining public input at the beginning of the process, the public becomes part of the mission. Planning assumptions, subordinate activities, and major milestone identification begins here.

Programming involves the segregation or collection of like activities, or activities requiring commonality of approach, into manageable parts. Further, programming identifies the context of the work relative to other activities. For example, a waste stream is the result of a production process. The process must be tied to the waste management activity in the programming phase. Also included in programming is the delineation of organizational responsibilities, authority, accountability, and funding. A journalistic approach works well in the programming phase. The manager must answer the top-level *who, what,* and *how* questions to add structure to the mission.

Baseline definition is the next important step. Picture the entire baseline resting on a three-legged stool. The legs are defined as the cost, technical, and schedule baselines. The stool, (i.e., the baseline), cannot support itself without the existence of all three legs (i.e., the baseline components). Second, a change of length in one leg requires readjustment of the other two. If the technical baseline changes, the "legs" of the schedule and the estimate must be revisited in order to balance the project once again.

At the program management stage the three baseline components are planned and documented. The actual content of the estimate, schedule, and technical pieces is defined later in the appropriate section. The three steps required to develop the framework of the baseline are the development of the (1) work breakdown structure (WBS), (2) the responsibility assignment matrix (RAM), and (3) the network logic diagrams.

The work breakdown structure is a breakdown of the components and subcomponents of a program such that each block represents an activity or collection of activities required for completion of the mission. Although the WBS looks like an organizational breakdown structure (OBS), the two are not the same. In fact, the most common error in developing a WBS is to have it follow an existing OBS. For example, if an environmental restoration activity involves transportation of waste drums, the OBS would show a contracted waste removal company (XYZ Co.), but the WBS may read "transport drums to receiving area No. 3 for final RCRA disposal."

By placing the OBS and the WBS next to each other, the RAM is developed. The RAM is a mapping from the OBS to the WBS and shows who is responsible for each work component. This important step will help identify any holes in the OBS that may require additional staffing or long lead time subcontractor procurement.

Network logic diagrams are used to begin sequencing the activities at the highest level. This is the first scheduling activity that should occur on a project, and it involves defining and ordering the activities required to accomplish the mission. The activities are shown on the newly created WBS, and are ordered based on the logic sequence necessary to complete the job. It is important that durations are not added to the WBS work elements at this time (i.e., concentrate on ordering the activities first).

Resource identification began with the development of the RAM. This matrix identifies the major resource disconnects between the mission and the current organization. For instance, in the earlier example, the XYZ Co. was contracted for waste transportation because no company resources existed to complete this activity. In the resource requirements step, the RAM is analyzed in detail. This step is especially crucial in environmental related programs because competition for these resources is high, and supply may be scarce in some area (e.g., environmental estimators). Keep in mind that resources are not just labor. Facilities, technology, equipment, and funds are all resources that must be addressed. The draft resource requirements plan is developed to detail the resource base, the resources and training needed, any procurement and acquisitions, and any research and development that needs to occur. This plan should be revisited as more is known about the project (e.g., during a dynamic baseline step), and should also be subject to document control.

As with any estimation of events that will occur in the future, predicting the outcome of environmental projects is not an exact science. As planning approaches reality, adjustments are made in the baseline to reflect these changes. The changes between an original baseline and subsequent updates are as important to document as the data contained in the new baselines. How, when, and why these updates occur become part of the history of the program.

4.2.2 Technical Scope Development

In estimating, a common problem exists when an original estimate is publicly announced. As changes to the scope occur, the tendency is to decouple the estimate from the scope. The owner will often remember the first number and, based on the new higher estimate as a result of scope change, will attribute the cost growth to "poor estimates" rather than the real culprit. Scope is the basis for the technical baseline, and should always be part of the estimating package when baseline dollar figures are reported. A schedule, with associated milestones, should also contain the scope elements when project goals are set. If this does not occur, the schedule may be met but the quantity of work to be completed will be in question.

Estimates and schedules must be balanced and based on a sound, technical scope document. Scope development is often not as rigorous as required. Because environmental projects contain much more narrative than blueprints or specifications, the need to communicate performance expectations fully is very similar to development of typical facilities' construction at a design criteria stage. The estimating teams on environmental restoration projects most often contain a program manager that knows the site and has a thorough understanding of the work required. The teaming of these individuals is one of the few ways to produce a good estimate since, as mentioned earlier, cost guides, databases and models in this subject area are rare.

The work scope must look at the site conditions, technology needs, regulatory issues, and facility and equipment requirements. The detail of the scope should also indicate what is *not* included, so as to avoid missing work elements (e.g., waste transportation to a burial ground). The support activities, such as safety, security, quality assurance, and occupational health, must also be addressed. Identification of all assumptions is important, and justification of each assumption should also be included. As more is known, these assumptions may be replaced by realities that may require a rebaseline if proven to be significant.

As a final note, the process of defining scope in the environmental arena requires familiarity with sites and site conditions, regulations, and available technologies that may be employed on the project. A strong team is required, and the estimators should be involved early on in the process.

Two of the primary areas in baseline development for environmental projects and programs are the overall program management and the development of scope. Because of the nature of this work, inadequate attention to these two areas will spell disaster for any estimate and schedule developed without this proper foundation. These projects are driven primarily by legal regulation, agreements, and consent orders and typically involve multiple agencies that must all acknowledge their role in the development of the baseline, and all agree on the work to be accomplished. As the field of environmental projects grows, project owners will require talent and expertise borrowed from the more traditional construction fields of estimating and project management.

4.2.3 Schedule Development

Schedules are an important facet of any project, and can be the primary focus of discussions and management decisions once the project is moving. This is especially true of environmental work. Time spent in the initial schedule development for the static baseline will be rewarded in fewer changes to the baseline during the project.

In most project scheduling scenarios the engineer, in consultation with the owner, develops a discrete set of scheduled tasks that coincide with the work elements in the WBS. This schedule is then embellished with factors that have a direct bearing on the plan such as procurement lead times and resource availability. Once fully planned, the schedule identifies the planned start and finish dates, and this becomes the static baseline schedule component.

In scheduling estimating projects, the development of this baseline component is critical and often very difficult. Remember the earlier definition of a baseline? The key words were that the baseline must be agreed to and signed off as the plan. Agreement on a schedule for environmental work is difficult to obtain primarily because of the number of the parties interested in the project. In order to please all advocates it is often necessary to plan the schedule and spend tremendous amounts of time incorporating comments from all parties. Although this is an untested formula (below), it illustrates the current thinking on development of environmental schedules.

Normal Construction
  Schedule planning: $I_n = D$
Environmental Construction
  Schedule planning: $I_n = D^x$
Where: I_n = Number of planning iterations
    D = Difficulty
    X = Number of owners or parties that sign off on the baseline.

Another aspect of scheduling that often occurs in environmental projects is the need to "backplan." Even in the earliest stages a date is often set in initial discussions for completion. The backplanning occurs when the end date is known and the tasks to get to the end date are identified. Then, as the network logic is developed, tasks are scheduled from the end point backwards to determine a start date that would allow the given end date to be met. A serious problem can occur if the end date is agreed to before the schedule is planned! Without the necessary planning, the danger is in agreements on consent orders that are unobtainable.

A final note on environmental schedules. Milestones are useful for tying an agreed-upon schedule to a contractual arrangement, such as a consent form or permit. Interim milestones are also helpful for measuring progress during the course of the project. Do not place undue emphasis on milestones as the focal points in the schedule, however. In practice, it is far more important to watch planned versus actual durations of activities as they progress because this will give you an earlier indicator of potential problems.

4.2.4 Estimate Development

Imagine a request to estimate a project whose only criteria statement was: "Build a hospital." Most estimators would pursue two fronts of data collection. First, a request for additional criteria would be made such as size, location, type, planned construction method, etc. The second avenue of information collection would be to research similar projects completed in the past and to look up any data collected on these types of projects in estimating manuals and handbooks. By pursing these two fronts, the estimator could produce a reasonable estimate, even at the conceptual level.

The project is now an environmental cleanup. The first avenue of pursuit is to locate additional information. If the assessment phase is not started, this site information is very difficult to obtain. The nature and extent of contaminants may only be estimated based on interviews of the history of the site. The second method of looking for data on comparable projects may be equally frustrating. Because the business of environmental cleanup is new, cost guides, books, and reference materials are scarce. Even when a comparable project is found, the likelihood of major regulatory changes between the time that the project was completed and the time of this estimate is significant. Depending on those changes, the old scope and associated project costs may be useless. The bottom line is that from a cost engineering perspective environmental estimates are a new, demanding area; developing an estimate is a major undertaking.

Estimate development is not possible at the preliminary stage, but the key is the recognition that updates will be needed quite often as more is learned from site investigation. A new, dynamic baseline update is often centered around a new estimate because the ideas on how to remediate a site can change drastically from the preliminary assessment/site investigation (PA/SI) stage to the completion of assessment. The first estimate must be designated as such, and all assumptions should be extensively documented.

Contingency and contingency analysis are important cost engineering tools in the development of environmental estimates for baselines (this topic is discussed further in Chapter 7). At the PA/SI stage, the scope is often an either/or remediation technique choice that has two or more options for cleanup that are equally likely. One way to deal with remediation technique choices with equal probability distributions is to provide separate estimates that can be combined to form a conceptual project range estimate. Also, as more becomes known, the options that become unlikely can be discarded. Although this is more work initially, it may present a more realistic project picture and can reduce cost growth surprises in the future.

If many projects of similar scope are planned, it may be cost effective to set up a project code of accounts (COA) for historical cost-gathering purposes. The added benefit of an environmental COA is the potential for comparison of high cost growth areas as dynamic baselines are developed.

The cost of a project is not a primary decision criteria for the selection of a remediation technique. Further, as mentioned earlier, the duration of the schedule and adherence to legal cleanup milestones force the cost of a project into secondary consideration. Both these conditions mean that more emphasis needs to be placed on initial estimate development because a reasonable amount of funding must be requested at an early stage in order for the project to continue.

4.3 DYNAMIC BASELINE MANAGEMENT AND CONTROL

So far in this chapter the focus has been on the development of the components of a credible static baseline (i.e., how do we come up with the starting point or snapshot in time?). Part of what makes this a difficult task, especially if a baseline is developed on a project/program that is under way, is that we quickly find out that a program/project is quite dynamic. Typically, a program/project goes through three stages of maturation that can be labeled as (1) conceptual, (2) design-oriented, and (3) definitive. Based on this evolution, the baselines must not only be developed, they also must be managed. To do so, a need exists for an integrated system of variance analysis and change order control put in place for tracking change after a baseline is developed. In addition, there must be a recognition that as a program/project matures, a comprehensive *rebaselining* effort is required, producing the dynamic baseline.

4.3.1 Managing the Technical Baseline

In hazardous waste management the technical baseline, or scope, is one of the three components required to fully define the program/project. As described earlier, the technical baseline is composed of the technical narrative (describing the content) and the assumptions used (describing the data shortfalls) in the development of the cost and schedule baselines.

The initial development of this baseline is through an iterative process. The technical baseline is developed as much as possible in advance of the estimate and schedule baselines. However, as these baselines are developed for the estimate and schedule (the other two required components), the technical baseline must be embellished with any additional information developed and assumptions used in preparing estimates and schedules.

Once the technical baseline is developed, it is managed in the same fashion that the cost and schedule baselines are, through an integrated system of change order control and variance analysis. These changes are then quantified in terms of cost and schedule. If the changes are the result of change orders, they are incorporated into the dynamic baseline. If the changes are true variances, they are reflected in the estimate at complete (EAC) cost calculation until the next stage of the program/project is reached. Based on the management practices in place on each specific project, the rebaselining effort considers the variances, and often the decision to incorporate or absorb these changes is made.

4.3.2   Managing the Estimate and Schedule Baselines

The differences between waste management (operations) and environmental restoration (assessment and cleanup) originate at the initial stages of the projects. Recall that, in environmental restoration (ER), projects have two phases known as assessment and remediation. The assessment phase (the first phase), which does not exist in waste management (WM), has two primary aspects of maturation, categorized as *conceptual* and *design-oriented*. Both ER and WM evolve as more data and information becomes known.

At the conceptual phase the estimates and schedules become the initial baselines for ER projects. These baselines are high-level in nature, and they attempt to define the major assessment activities and their associated costs and schedules.

Once these baselines have been established at the conceptual stage, a contractor is typically retained for the assessment phase. The contractor has the responsibility to become familiar with the conceptual baselines that were developed. In addition, the contractor and client must establish a workable system of tracking variances and change orders. This system will allow the contractor to process change orders that typically allow relief in contractual terms of cost and schedule. If there are variances, they can be identified by the contractor as they occur; both the contractor and the client have immediate insight into the potential growth in the estimate-at-complete benchmark. Finally, the requirements for the remaining estimate and schedule baselines need to be agreed upon so that the level of detail and the elements of the estimate and schedule baselines can properly be developed, maintained, and managed.

The agreed-upon time at which the assessment phase matures to the design stage is not a fixed point, and typically occurs at a 20% to 30% complete milestone in the work plan. The contractor and client can feel comfortable that the information available is sufficient for the rebaselining effort at that time. This stage will last from the Work Plan through to the end of the assessment phase, and culminate in a Record of Decision (ROD). These design baselines are based on a formal work plan. This estimate is expected to be unit price in nature because of the maturity of the information and level of detail provided in the investigative phase of the ER project. It is also very important that a comprehensive contingency analysis be done so that the total estimated cost will be within the agreed-upon level of reliability. In addition, the schedule is expected to show all the sub-activities. The baseline will be equal to the sum of the conceptual static baseline plus variances and change orders to date. This then establishes the design baseline. Without variance analysis and change order systems in place, this detailed baseline update will not reflect the project history.

Near the end of the assessment phase there should be a conceptual level baseline developed for the remediation phase of the ER project. This baseline should meet the criteria of a typical conceptual baseline for design and construction.

From this point of ER remediation, the waste management baseline looks very similar. To a large extent data is fairly well defined; the baseline is being used to manage the hazardous waste removal, treatment, storage, or disposal; and further baselining will occur at a detailed design stage or when significant changes are encountered.

As the contractor establishes a detailed design-stage baseline, the contractor will have specifications, quotes, and delivery schedules of the major components (engineered equipment), which will be incorporated into the baseline. For both the ER and WM program/projects, associated bulk material quantities needed are also verified. These baselines will be the sum of the conceptual stage baselines plus the variances and change orders up to this point in the program/project.

The final stage of program/project maturation is the definitive stage. At that point, completed drawings and specifications are available. Based on this definitive data, a baseline is developed for final construction. Field progress, change orders, and variances will be managed relative to this set of baselines. This set of variances and change orders will provide the final link of the overall program/project history. Based on this final baseline and previous historical baselines, a closeout report will be developed. This information and history will be important in the evaluation of future environmental restoration projects and waste management programs.

4.4 CONCLUSION

Although the cost and time required for development of an effective baseline may seem like an unnecessary expense, its advantages far outweigh its disadvantages. The advantages can be identified from both the original static perspective and the dynamic management perspective. From the static viewpoint, the ability to completely develop the cost and associated schedule for a project produces better preliminary project communication and management decisionmaking. Further, as part of this development the contractor should be able to identify and assess the areas of uncertainty for the project and communicate the perceived risks. This helps to prevent or minimize "surprises" to the client and regulators, and allows for a defense of the project costs and schedule. Considering the controversial and legal nature of a hazardous waste project, doing so can result in considerable savings of both cost and time. From the dynamic perspective, the effective baseline becomes a management tool that allows for early identification of potential cost, scope, and schedule overruns. It also allows for the identification and quantification of cost and schedule changes caused by outside influences. In addition, it provides a complete path or history of the program for both the contractor and client and can be used as part of a "lessons learned" or self-assessment tool for the next project.

Producing a baseline and managing the project with a complete document set of updates baselines is particularly useful for hazardous waste projects. As more hazardous waste projects are completed over the next few years, this overall data set will become the basis for smarter planning and more precise regulation and budget control.

5
Estimating the Remediation of Hazardous Waste Sites

Ronald G. Stillman
Roy F. Weston, Inc., West Chester, Pennsylvania

5.1 INTRODUCTION

The remediation of a hazardous waste site is a difficult operation that combines construction expertise with unique scientific and engineering requirements. The estimating of the work is equally arduous; the site conditions, environmental restrictions, and the construction approach present the estimator with problems and applications that must be carefully planned and costed. Remediation work involves approaches and requirements that are only applicable to this type of work.

An experienced estimator will apply normal estimating techniques when developing the costs. There are hidden costs associated with remediation work that have to be accounted for by allowances and assumptions. Unlike general construction, estimating manuals are not available with unit prices for remediation work.

This chapter will address the following areas with details and examples:

- Estimating techniques
- Development of estimates for three alternative soil remediation projects
- Hidden costs of remediation work

The hazardous waste remediation estimate is used for many purposes. The consultant will use the estimate during the feasibility study to assist in determining the preferred method of remediation. At the completion of the detailed design, the estimate is used for funding purposes and as a comparison to the bids that are received for the work. The contractor bidding on the work develops an estimate as part of the proposal. The techniques described in the following subsection can be used for all estimates.

5.2 ESTIMATING TECHNIQUES

5.2.1 Pre-estimate Activities

The initial work in the development of a hazardous waste estimate is similar to that of general construction with the same procedures and steps followed. Some of these are:

• Work plan and work breakdown structure (WBS) development
• Identification of direct hire and subcontract work
• Determination of local conditions

The development of a work plan is necessary prior to starting the estimate. This plan details the sequencing of the work activities for the project. Each of the activities is defined and any relationship between them is identified. The crews and equipment that are required for each activity are based on the plan. This document provides the basis for both the project estimate and the schedule. The work plan is based on a scope of work for the project which can be the record of decision (ROD) for the project, an early project specification, or a detailed bid document.

The WBS for the project is based on the work plan and is the basis for the estimating account structure. The WBS provides the link between the estimate and the schedule (and accounting system in the execution of the work).

The local conditions affecting cost and schedule must be researched prior to starting the estimate. The location of the project has a major impact on the cost and schedule. If the site is located in an urban or residential area, safety and security precautions are more stringent than those for a rural area. Work hours may be curtailed due to noise or congestion at the urban site. A rural site has its own potential cost impacts such as lack of a work force, lack of utility access, and the scarcity of material suppliers and temporary living facilities.

The forecast of work in the region of the site must be reviewed for the time of project. The work forecast will identify the impacts to the estimate of the following:

- Available work force
- Bidding climate
- Construction equipment and tools availability

If the work forecast is heavy, the contractors bidding on the work will not conduct the bidding process aggressively enough. The resulting high overheads and fees will be factored into the estimate. There may not be an adequate work force available for the project, and premiums such as overtime or bonuses may be required to attract workers. Construction equipment and tools may not be available, or may be available only at a premium. Conversely, if the work forecast is weak, the contractors will develop an aggressive bid with low overheads and fees. A work force can be attracted without premiums. Construction equipment and tools may be available at discounts. The local area may also determine whether the work will be performed by union or merit shop contractors. A checklist of local conditions that the estimator must review and complete prior to developing the detail costing is included as Table 5.1.

Following the site surveys and pre-estimate analysis, the estimator begins the detail costing by the sequences of the project such as:

- Premobilization activities
- Mobilization including temporary facilities and utilities
- Craft labor analysis
- Health and safety impacts
- Special materials and equipment
- Transportation and disposal
- Demobilization

5.2.2 Premobilization Activities

The premobilization period is important in the remediation project. Work must be completed in order to start on site. All project-related activities are charged against the work scope. The general premobilization activities include plan writing, home office support including procurement and project controls, relocations, and worker training and physicals. All costs that are required to begin work must be included in the estimate.

After receiving the notice to proceed on the project, the contractor will begin writing the contract-specified plans, which can include the following:

Table 5.1 Site Checklist

1. Accessibility to site
   a. Rail
   b. Barge
   c. Roads (any bridge limits)
   d. On-site roads

2. Utilities on-site
   a. Existing or brought to site
   b. Type of utilities: gas, electric, water
   c. Voltage of electric power
   d. Names of utilities' companies

3. Site conditions
   a. Topography
   b. Site roads
   c. Congestion
   d. Access limitations
   e. Health and safety level required
   f. Water problems/table
   g. Type of soil
   h. Rock

4. Labor force
   a. Union/open shop
   b. Availability (unemployment in area)
   c. Proximity to town (size)
   d. Trailer park required?
   e. Minority requirements
   f. History of labor in area

5. Material supply
   a. Proximity to supply houses
   b. Type of supplies required
   c. Worked with firms before

6. Subcontractors
   a. Client preferred
   b. Experience with any in area
   c. Sope of subcontractor work
   d. Merit shop/open shop/union

7. Equipment
   a. Construction equipment firms in area
   b. Supply houses for tools

Table 5.1 (*continued*)

  c. Delivery constraints for engineered
     equipment due to bridges, road lim-
     itations, barge facilities nearby, rail-
     road points

8. Miscellaneous activities
  a. Local opposition/support
  b. Client interest
  c. Competition
  d. Weather-related conditions/flood-
     plains, etc.

- Quality assurance and quality control plans
- Site health and safety plan
- Work plans
- Environmental risk and protection plans

The personnel time required by the contractor to write the plans is estimated, including trips to review the plans with the regulatory agencies
and publications. Failure by the contractor to provide adequate staff
(and budget) for this work can impact the start of the project work at the
site. General home office support costs for a project such as procurement,
project controls, and management needs to be included in the estimate.
The home office involvement for hazardous waste projects is usually
greater than that for general construction.

An important step during the premobilization period is to obtain a
trained labor force for the on-site work. All site personnel must be trained
in accordance with OSHA standard 29 CFR 1910.120. This training consists of an initial 40 hours of classwork and then an 8-hour class each year
to maintain the certificate. The on-site personnel also must be physically
fit to work at the site and must pass an annual medical examination.

A contractor will charge costs associated with the OSHA requirements
to the project rather than as part of the firm's overhead. The costs include the labor hours, medical examinations, and training courses. A
turnover rate will be assumed in calculating the estimated number of workers that will be examined and trained for the project.

5.2.3 Mobilization

The on-site mobilization for the project has steps that must be accomplished prior to the remediation. Their costs must also be included in the estimate. These steps may include:

• Installation of temporary construction facilities
• Installation of personnel and vehicle decontamination plants
• On-site environmental controls
• Implementation of special long-term remediation systems such as an incinerator or water treatment system

The first activity in mobilization is the setting up of temporary construction facilities, which are similar to those required in general construction. The facilities are erected in the clean area of the site and therefore do not have hazardous waste impacts associated with them. Key cost items include layout of the site, layout of trailers, utility connections to the site, and the construction of roads and parking areas. An area is usually designated and developed for the storage of hazardous waste that will be transported off-site.

The decontamination facilities are erected in the area that is designated the contamination reduction zone. However, when they are installed, the area is clean. These facilities can include trailers, the equipment decontamination pad, the collection and storage system for the rinse water, and personnel decontamination pad. The project health and safety plan will define the decontamination requirements for the project.

Environmental monitors and controls, which are established on site during mobilization, are defined in the project environmental control and health and safety plans. The types of monitoring required will include air and water monitors, erosion controls, and a meteorological station. Costs to install and to test the site conditions initially are included in the estimate.

The mobilization of a long-term remediation system such as a transportable incinerator or a water treatment system entails the construction and start-up of the unit.

Prior to beginning the remediation work on site, all on-site workers (both supervisory and manual) will undergo site-specific medical examinations and training. The medical examination may consist of a blood test. The on-site training can last several days and may recur whenever each new phase of the project is begun.

5.2.4 Craft Labor Impacts

Craft labor is a major cost component of a remediation estimate. Impacts to the labor exist that can cause the cost to increase; these include:

• Wearing of personal protective equipment (PPE)
• Productivity loss in the execution of the work
• Wage increases
• The requirement for additional personnel

When developing an estimate, the impact to the cost for each of these items must be reviewed carefully.

The PPE that the laborers must wear for each phase of the remediation is defined in the project health and safety plan. Each phase of the work may have different PPE requirements. Table 5.2 lists the protective equipment that the worker must wear during the performance of the work at the different levels.

Table 5.3 presents an example of developing the PPE costs for a modified Level D project. This example assumes one change of outer garments per day for each worker. Impacts that cause variances in the calculations of cost per person per day or cost per hour include:

• Number of outfit changes per day
• Length of work day
• Changes in PPE requirements during the work day

In developing the detail estimate, the labor hours are summarized by the length of time that the workers spend in each level of protection. The

Table 5.2 PPE Requirements for Safety Levels

Level of safety	Equipment required
D Modified	One-piece coverall, disposable outer boots, steel-toe boots, Tyvek overall, boot covers, inner gloves (surgical), outer gloves, hard hat with face shield.
C	Same as D modified, except with a respirator with cartridge.
B	Same as D modified, except with a self-contained breathing apparatus.
A	Same as B, except with a fully encapsulating suit.

Table 5.3 PPE Cost Development for a Modified Level
D Project

Equipment required quantity per day/person[a]	Useful life	Cost per day ($)
Tyvek overall (2 each)	Once	6.00
Boot covers (2 pair)	Once	10.00
Inner gloves (2 pair)	Once	1.00
Outer gloves (2 pair)	Once	10.00
Work boots (1 pair)	6 months	1.00
Hard hat (1 each)	3 months	0.50
Cotton coverall (1 each)	3 months	0.50

[a]Total cost per person per day = $29.00; Total cost per hour (8 hrs/
day) = $3.62.

hourly costs of PPE are developed for each PPE required for the project.
The total hours for each level of protection are then multiplied by the
cost per hour for wearing PPE. An example of Level D Modified (assuming 5,000 hours of Level D Modified work) is:

Level D Modified: 5,000 hours × $3.62/hour = $18,100

The costs for the ultimate disposal of the contaminated PPE are to be
included in the estimate.

As part of the estimating process, a productivity analysis for craft labor
on hazardous waste projects is performed. The analysis is similar to that
performed on general cosntruction projects but includes additional impacts such as time lost due to dressing and undressing, weather, and loss
of dexterity. Chapter 6 discusses the impacts on labor productivity.

A productivity analysis for craft labor is shown in Table 5.4. Only the
four more common levels of protection are analyzed; these are normal,
Level B, Level C, and Level D Modified. Each level of protection is analyzed
for the work activities that impact craft labor. The driving factors to the
loss of productivity are described in Chapter 6. Each project will have
unique productivity adjustments based on the conditions presented in the
work scoped. It is not unusual for distinct activities of a remediation,
such as demolition, excavation, and capping, to have different productivity adjustment factors.

Table 5.4 Productivity Adjustments

Activity	Normal conditions (minutes)	Level B conditions (minutes)	Level C conditions (minutes)	Modified Level D conditions (minutes)
Lost time				
Wait for instructions	30	30	30	30
Travel time	15	15	15	15
Personal time	40	40	40	40
Total, lost time	85	85	85	85
Support time				
Pick up tools	30	30	30	30
Measurement	10	10	10	10
Outfitting	0	120	120	30
Recovering	0	60	40	10
Total support time	40	220	200	80
Direct time on job				
Time =	355	175	195	315
Loss due to suits	0.00%	60.00%	30.00%	10.00%
Loss of efficiency	0	105	59	32
Total available work time	355	70	136.5	283.5
Total, daily (8 hrs)	480	480	480	480
Percent work/total time	73.96%	14.58%	28.44%	59.06%
Adjustment factor to Normal conditions (Excludes support craft)	Base	5.07	2.60	1.25

As shown in Table 5.4, the *lost time* category includes instructions, travel time to the work area, and personal time (lavatory breaks). In the example, 85 minutes are assumed to be lost on a daily basis.

The second activity is *support time*, which consists of tool management, measurements, and remediation activities such as outfitting and recovery time. The outfitting time is the time it takes for the worker wearing PPE to dress and undress. The amount of time required for this activity depends on the number of changes that are required each day and the level of protection.

At the end of a shift or the allowable work duration, the worker must recover from exertion. The recovery time is dependent on the work assignment, the temperature of the work area, and the PPE worn. As shown in Table 3.5, support time can vary from the normal conditions' length of 40 minutes to the Level B duration of 220 minutes.

The amount of time available for the laborer to work on the project is called *direct time* and is calculated by subtracting the lost time and support time durations from the total work day duration. An example for Level C follows:

Total work day = 480 minutes (assumes 8-hour day)
Lost time   = 85 minutes
Support time   = 200 minutes
Direct time   = 195 minutes

The direct time on the project is then modified because of the loss in efficiency as a result of wearing the PPE. Included in this efficiency calculation are impacted caused by loss of dexterity and temperature extremes and additional health and safety safeguards. For Level C, the efficiency loss is estimated at 30%, or another 59 minutes lost per day. In this example, the actual time the laborer performs work on the project is forecast to be 136 minutes or about 28% of the total day.

The development of productivity factors is subjective and is unique for each project. Based on the cost drivers described in Chapter 6, factors for each project are developed. Different factors can be calculated for distinct portions and activities within a project also. The productivity adjustment to normal conditions for each level of protection is calculated as follows:

Productivity adjustment = % work PPE/% work normal

As shown in Table 5.4, for Level C the productivity adjustment is equal to 73.96%/28.44%, or 2.60.

For each work activity, the normal construction productivity unit must decrease by 2.6. (It takes a person in Level C about 2.6 times as long to perform a task as compared to normal conditions.) Due to the low productivity on the project and the lost time dressing and undressing, the construction manager must attempt to maximize the work time available. One approach is to work overtime. The dressing and undressing times occur before or after the work day. The craft worker is usually paid a premium for this overtime. A second wage rate impact is an hourly bonus (or wage rate increases) paid to craft labor for working in respirators and self-contained breathing apparatus (i.e., Levels C, B, and A work). The additional hourly increase can range from 10% to 20% of the base wage rate.

A remediation project requires more personnel than a general construction project. These additional personnel are both supervisory and craft labor. The supervisory staff for a remediation project requires a site health and safety officer and a quality assurance engineer. The site health and safety officer is responsible for ensuring that the remediation effort follows the project health and safety plan. This person is independent of the site management to help ensure that the project will be conducted according to plan. The quality assurance engineer verifies that the work is being performed as specified and is responsible for ensuring the integrity of the sampling effort.

The craft labor additional personnel include:

- Sampling technicians to monitor the work. They will obtain all the samples required for the project and ship the samples to the laboratory for analysis.
- At least one full-time person to operate the equipment and personnel decontamination pads.
- Safety watch personnel who observe the craft laborers working at the site. These persons will notify the supervisory staff when an accident occurs. Depending on the activity, the ratio of safety watch personnel to the workers may be as low as one to one.

The length of time a laborer can work while wearing PPE is limited. For the remediation to operate continuously throughout the workday, a second or third shift of workers may be required. While one shift is working, the other shift will be recovering or performing work in the clean areas. Because of this shiftwork, more laborers are required in the work force.

5.2.5 Sampling and Analysis

An important activity during a remediation project is the sampling and the follow-up analysis. These activities are safeguards for the on-site personnel and the surrounding environment. Typical sampling activities include perimeter monitoring, real-time sampling in the work area, and final verification of remediation. Environmental monitoring stations are located around the perimeter of the site to monitor the air and water so as to verify that contaminants are not migrating off the site. Samples are taken from these stations on a periodic basis and are one of the responsibilities of the sampling technicians. As work progresses, the contaminant levels in the work areas change as areas are cleaned or exposed. The sampling technicians are required to be present during the different phases of the work to monitor the levels. Changes to the PPE will result as the readings vary. If the levels rise too high, work may be dlayed or the area may have to be evacuated.

At the completion phase of the remediation, samples are taken from the cleaned areas to verify that the designated level of concentration has been met. For example, an area being excavated to the prescribed depth will be sampled in several places. Samples will be analyzed before the area is declared clean and backfilling is done. The time required for the analysis may be a few days to several weeks. The crew working in the area may need to be moved to another area or demobilized off site if the analytical delay is lengthy; otherwise, the crew can remain at the site. This delay between completing the work and receiving the final analysis verifying the closure must be accounted for in the estimate by the demobilization and remobilization of the work crews or in the nonwork period costs for the crew. The cost of each sample analysis varies from $25 to over $1,000 depending on the type of analysis required and the turnaround time for the results. In developing the cost of analysis, the estimator must know the quantity of samples that will be required, the analysis for each, and the time in which the results will be needed.

5.2.6 Special Materials and Equipment

Many common materials cannot be used for handling hazardous materials because of chemical reactions with the contaminants. An analysis of the contaminants is completed before work is started on site in order to mitigate any problems. A typical occurrence may be loading a rubber-lined tank truck with a liquid that can react with the lining. As part of the estimate, the estimator must review the proper handling procedures for the contaminants. Exotic materials may be required for the project. Both the time it takes to obtain the materials and the higher cost of the materials must be accounted for in the estimate.

The equipment used for remediation has been adapted from general construction equipment (such as backhoes and forklift trucks). The equipment may have been modified with an enclosed cab having its own supply of air and a blast shield. The cost of renting or leasing this specialty equipment is higher than for general construction equipment. As the work progresses the equipment is decontaminated as it leaves the site, at which time parts may be damaged, requiring replacement. For some projects the piece of equipment may not be salvageable and must be disposed of as hazardous materials. The cost of equipment maintenance and repairs should be included in the estimate.

5.2.7 Transportation and Disposal

The transporting and disposal of hazardous waste is an important component of a remediation project and can range to 80% of the entire cost of the remediation. The estimate for this activity is therefore critical in the development of the project cost. Typical items that are disposed of from a remediation include:

- Contaminated materials such as soil and concrete block that is being excavated or removed
- Liquid pumped from the ground
- Decontamination liquid from cleaning vehicles, equipment, and personnel
- PPE

The ultimate method of disposal for the contaminated material is regulated by the government depending on the contaminants involved. Types of disposal may include:

- Landfilling
- Incineration
- Deep well injection
- Chemical stabilization

The costs for each type of disposal vary depending on the quantity and the contaminants.

The transportation of the materials has to be planned in advance of the on-site work and may impose schedule impacts. The material cannot be transported to the disposal facility until a laboratory analysis of the contaminated material is performed. Samples are taken at designated intervals based on time or quantity. The time for the analysis may delay the shipment. In some instances, the contaminated material is stockpiled on site for months while the analysis is being performed and the disposal facility accepts the waste. Costs for the demurrage of the storage containers are to be included in the estimate. The field crews may have to be remobilized to load the transporting equipment such as trucks or railroad cars.

The cost of transportation and disposal for materials may include:

- Hauling costs to the disposal facility
- Demurrage costs for the material containers

- Sampling and analysis costs
- Stabilization and disposal costs including surcharges
- Crew demobilization and mobilization costs

5.2.8 Demobilization

The demobilization for a remediation project is similar to that for general construction except for the decontamination procedures that are required. The major steps in the demobilization of a hazardous waste project are:

- Securing of the site
- Disposal of the contaminated materials
- Removal of the decontamination facilities
- Removal of the construction facilities
- Final medical tests for the on-site workers

The securing of the site includes fencing in of the site, placement of monitor wells, and the posting of signs around the perimeter of the site. The removal of the decontamination and temporary construction facilities occurs during this final phase. The decontamination facilities may have to be demolished and the debris transported to a hazardous waste disposal site. Temporary construction facilities such as trailers are hauled from the site. The roads and parking area required for the project are removed. The paving material can be considered hazardous and may be transported to a hazardous waste site. When they leave the site, the laborers are often required to undergo medical tests to determine if they have absorbed contamination from working on the site. Costs must be included both for the medical tests and the labor time involved in the testing. The off-site demobilization will include the final disposition of the waste handling equipment and the decontamination systems.

5.3   EXAMPLES OF PROJECTS

5.3.1 Alternatives

There are three common approaches for the remediation of a landfill or contaminated area:

- Off-site excavation and hauling of the contaminated material
- Impermeable capping of the site
- On-site incineration of the contaminated material

The method chosen depends on various items including the estimated costs, local involvement, and the contaminants. Studies are performed to select the proper technology for the remediation. Other potential options for a soil remediation project are bioremediation and in situ volatilization.

Estimates for each of the three common alternatives are included in the appendixes at the end of this chapter. The costs developed reflect 1992 prices. General assumptions include:

- Demolition will not be required.
- Area of the site = 300,000 ft^2 (about 7 acres).
- Area of the contamination = 170,000 ft^2 (about 4 acres).
- Depth of contamination = 4 ft average.
- Approximately 25,000 yd^3 (35,000 tons) of contaminated soil.
- Suburban location with houses nearby.
- Groundwater is below excavation depth.
- Excavation will be Level C.
- Backfilling will be Level D Modified.
- Utilities to the perimeter of the site.
- Average wage rate = $30/hr with a $2 premium for Level C.

The mobilization activities and productivity analysis is similar for all examples.

5.3.2 Excavation and Hauling Alternatives

During this approach, all contaminated soil will be excavated and placed into 20-yd^3 roll-off containers. Runoff water will be collected and disposed of off site. Based on a productivity analysis, the excavation production in Level C is expected to take about twice as long as general construction. The backfilling has a 20% productivity penalty associated with the operation. For the work, the equipment required and production is as follows (each piece has an operator associated with it):

- Bulldozers—three each
- Front-end loaders—two each
- 20-ton dump trucks—six each
- Excavation production—1,000 yd^3/day

A decontamination person, a flagman/traffic director, and two laborers are assigned to the work force. The laborers are required to assist in loading the roll-offs. The estimated cost for this approach is approximately

$20.8 million and the on-site work is expected to last 6 months with the off-site transport of the soil occurring during that time frame.

5.3.3 Impermeable Capping

The impermeable capping of the site is based on a membrane liner system. The procedure for this work consists of the following activities:

- The contaminated area is cleared and grubbed.
- The area is the graded with consolidation of the contaminated material.
- A layer of sand is placed over the area to be lined.
- The layers of the membrane lining are placed down with backfill as required.
- The area is backfilled with soil and topsoil and then seeded.
- Monitoring wells are installed on the perimeter of the site to verify that contaminants are not affecting the groundwater.

The estimated cost for this approach is $3.1 million, and the forecast construction duration is 6 months.

5.3.4 On-site Incineration

The incineration of the on-site material using a high temperature transportable incinerator system is a more lengthy process of remediation. The incineration systems currently used on projects have a nominal burn rate ranging from 3 tons/hr to over 20 tons/hr. The estimate will assume a "mid-range" system capable of incinerating 8 tons/hr. An on-line factor of 65% is estimated to allow a project-specific incineration rate of about 5 tons/hr. The unit will operate 24 hours a day, 7 days a week. Based on these factors, the 35,000 tons of material will be incinerated in 300 days. Each day, three shifts of operators will be required to operate the unit. Overall, 25 persons will be required daily at the unit, including maintenance personnel. The average shift will last 9.5 hours due to dressing/undressing time and shift overlap requirements. The estimate is based on 10 hours daily per person to cover the premium portion of overtime that is required to be paid. Other daily costs for the incinerator include chemical, utilities, miscellaneous expenses, and laboratory, as well as allowances for depreciation and major maintenance of the system.

Typical chemicals consumed during the unit's operation include caustic, carbon, oxygen, and calibration gases. The daily costs for these chemicals are project-specific because of differing reactions with the contaminants.

An allowance of $700 per day during operations is assumed for the estimate. The types and quantities of utilities consumed during the operations are based on the incinerator design and the composite of the contaminants being incinerated. Most incinerators use either electricity or gas to produce the high temperatures required for operation. In this example, a gas-fired incinerator is assumed. The composite of the contaminant (i.e., the moisture content, type of material, and heating value of the material) directly influences the utility usage. A daily cost of $2,800 is assumed for the estimate.

Miscellaneous expenses include all other operating costs incurred. These costs include equipment rental (backhoes and loaders), health and safety equipment, supplies, and consumables, office and decontamination trailers, office expenses such as janitorial and copying, home office travel to the site, and incidental repairs. For the example, the estimated daily cost for miscellaneous costs is $4,000. The operating cost for an incinerator system is the establishment of a fund for any major maintenance that will be required, such as refractory lining repair. An allowance for depreciation should be included in the operation costs for the unit. Funding for these items can be part of the overhead for the company, or can be charged directly to the unit. In this example, maintenance and depreciation have been included at a daily allowance of $2,000.

The procedure for installing the transportable incinerator system on site is as follows:

- The system is transported onto the site and erected on concrete foundations.
- The system is "shaken down" to verify and correct operating procedures during the test and start procedure.
- The system starts to incinerate the contaminants and is tested by the Environmental Protection Agency (EPA) for stack emissions (test burn).
- The incinerator cannot operate at full capacity until EPA approval of the test burn is received (a 3-month duration is assumed).
- If the system has performed acceptably in the past, the incinerator may operate at a reduced capacity (interim operations) during the approval time.
- Upon receipt of the EPA approval of the test burn, the system will operate at full capacity.

There must be sufficient space available on the site to store soil until it can be incinerated. Depending on the size of the area, the excavation routine will be determined. If the storage area is large enough, all the

Table 5.5   Cost and Schedule Comparison of the Three Alternatives

Alternative	Construction cost	Remediation duration
Excavate and haul off-site	$20.8 million	6 months
Liner system	$ 3.1 million	6 months
On-site incineration	$18.7 million	15 months

contaminated soil may be excavated and stored. If only a small area is available, then the contractor will only excavate what that area can accommodate.

After the incineration, the soil ash is still considered a hazardous waste and is placed in a secured landfill (unless testing of the ash proves that all contaminants are removed). The area that is excavated can be used or the soil is transported off-site to a secured landfill. The example assumes that the incinerated soil ash will be backfilled into the excavated areas and the area capped with a membrane liner system. The estimated cost for the incineration alternative is $18.7 million and the work duration is 15 months.

5.3.5 Comparison

Table 5.5 is a cost and schedule comparison of the three alternatives. The selection of the remediation method for a site is not based solely on cost or schedule, but also on local needs and the long-term effects of each alternative. The liner system is the least costly alternative; however, this approach does not remove the contaminants. The site is environmentally safe, but the contaminants still exist.

5.4   HIDDEN COSTS OF REMEDIATION

5.4.1 Hidden Costs

Remediation work has hidden cost impacts associated with the performance of the work. Some of these are:

- Unique schedule upsets
- Remediation techniques
- Bonding restrictions and costs

The estimator has to review these potential impacts and include cost allowances as necessary. See also Chapter 4 for additional hiddens costs of the work.

5.4.2 Schedule Upsets

The remediation of a hazardous waste site is under the scrutiny of several groups including state and federal regulatory agencies, the environmental consultant, the construction manager, and local citizen action groups. Many of these organizations review and approve the submittals and oversee the work. Due to the large number of participants, delays in the approval of the submittals may occur or work restrictions may be enacted at the site.

5.4.3 Remediation Techniques

The technique proposed for the remediation work may be proprietary to a firm, or a technique may be supplied by only a few firms, which is the case for a transportable incinerator. The estimator developing a budget estimate must be cognizant of the market conditions for the remediation technique and adjust the overhead and profit margins in the estimate accordingly. The unique techniques may limit the competition on many sites.

The selected technique may be difficult to perform. The constructibility of the project must be reviewed by the estimator. Additional costs may result becuase of the creative nature of the remediation.

5.4.4 Bonding

Currently, obtaining a performance and payment bond for hazardous waste work is more costly and more restrictive than obtaining one for general construction. The bonding limits placed on the company for a remediation project may be lower than for other project types. This bonding restriction may limit the number of competitors for projects. The estimator must be cognizant of the limited competition as well as the higher cost for the bond in the development of project estimate.

5.5 CONCLUSION

During the development of the estimate, the estimator must consider the many unique aspects of a remediation project. The major concerns are the protection of the site laborers and the surrounding environment. By following the steps described in this chapter, an estimator will be able to develop the costs for remediating a hazardous waste site.

Appendix 5A: Example of Conceptual Estimate Excavation and Transport
Offsite Approach

DESCRIPTION	QUANTITY	HOURS	LABOR RATE	TOTAL LABOR	MATERIAL UNIT	TOTAL	SUBCONTRACTED UNIT	TOTAL	TOTAL
SUMMARY									
TOTAL, PREMOBILIZATION ACTIVITIES				$29,560		$615,650		$0	$645,210
TOTAL, MOBILIZATION				$12,414		$10,550		$40,050	$63,014
TOTAL, REMEDIATION WORK				$149,520		$310,500		$14,583,750	$15,043,770
TOTAL, INDIRECTS				$144,000		$94,622		$0	$238,622
TOTAL, DEMOBILIZATION				$6,750		$500		$33,400	$40,650
SUBTOTAL				$342,244		$1,031,822		$14,657,200	$16,031,266
CONTINGENCY	20.00%								$3,206,253
FEE	10.00%								$1,603,127
TOTAL, ESTIMATE				$342,244		$1,031,822		$14,657,200	$20,840,646
PREMOBILIZATION ACTIVITIES									
WRITING OF PLANS									
SITE HEALTH & SAFETY	1 LS	80	$40.00	$3,200					$3,200
WORK PLAN	1 LS	80	$40.00	$3,200					$3,200
QUALITY ASSURANCE	1 LS	60	$40.00	$2,400					$2,400
AIR MONITORING	1 LS	60	$40.00	$2,400					$2,400
SITE SECURITY	1 LS	20	$40.00	$800					$800
OBTAIN LOCAL PERMITS	1 LS	80	$40.00	$3,200					$3,200
COPYING/PUBLICATIONS	1 LS				$2,000	$2,000			$2,000
TOTAL, WRITING OF PLANS				$15,200		$2,000		$0	$17,200
HOME OFFICE SERVICES									
PROCUREMENT	1 LS	80	$30.00	$2,400					$2,400
PROJECT CONTROLS	1 LS	40	$30.00	$1,200					$1,200
GENERAL MANAGEMENT	1 LS	20	$50.00	$1,000					$1,000
TOTAL, HOME OFFICE SERVICES				$4,600		$0		$0	$4,600
CFR 1910.120 MEDICAL/TRAINING									
MEDICAL EXAMS	12 EA	48	$20.00	$960	$500	$6,000			$6,960
40 HR TRAINING	4 EA	160	$20.00	$3,200	$600	$2,400			$5,600
8 HR REFRESHER	5 EA	40	$20.00	$800	$250	$1,250			$2,050
TOTAL, CFR 1910.120				$4,960		$9,650		$0	$14,610
PRECONSTRUCTION MEETINGS									
PROJECT MANAGER	1 LS	32	$50.00	$1,600					$1,600
PROJECT ENGINEER	1 LS	48	$40.00	$1,920					$1,920
SITE MANAGER	1 LS	32	$40.00	$1,280					$1,280
ALLOW FOR TRAVEL	1 LS				$4,000	$4,000			$4,000
TOTAL, PRECONSTRUCTION MEETINGS				$4,800		$4,000		$0	$8,800
PERFORMANCE & PAYMENT BOND									
ALLOWANCE FOR	1 LS					$600,000			$600,000
TOTAL, PREMOBILIZATION ACTIVITIES				$29,560		$615,650		$0	$645,210

Appendix 5A: *(continued)*

DESCRIPTION	QUANTITY	HOURS	LABOR RATE	TOTAL LABOR	MATERIAL UNIT	MATERIAL TOTAL	SUBCONTRACTED UNIT	SUBCONTRACTED TOTAL	TOTAL
GENERAL MOBILIZATION ACTIVITIES									
MOVE IN + SET UP TRAILERS									
OFFICE	1 EA	24	$30.00	$720	$250	$250	$250	$250	$1,220
DECONTAMINATION	1 EA	48	$30.00	$1,440	$300	$300	$500	$500	$2,240
TOOL/EQUIPMENT	1 EA	8	$30.00	$240	$150	$150	$100	$100	$490
TOTAL, MOVE IN TRAILERS				$2,400		$700		$850	$3,950
MOVE IN EQUIPMENT									
BULLDOZERS	3 EA						$250	$750	$750
DUMP & WATER TRUCKS	7 EA						$100	$700	$700
FRONT END LOADERS	2 EA						$250	$500	$500
ROLLER COMPACTOR	2 EA						$250	$500	$500
TOTAL, MOVE EQUIPMENT				$0		$0		$2,450	$2,450
SET UP WORK ZONES + INITIAL SITEWORK									
SURVEYOR	2 DAY						$650	$1,300	$1,300
TAPES AND STAKES	1 LS				$150	$150			$150
CLEARING + GRUBBING	1 LS	36	$32.00	$1,152	$250	$250	$250	$250	$1,652
GRADING	1 LS	36	$32.00	$1,152	$250	$250	$250	$250	$1,652
GRAVEL PAVING	200 CY	40	$30.00	$1,200	$12	$2,400	$0	$0	$3,600
ELECTRIC CONNECTIONS	1 LS	24	$30.00	$720	$1,000	$1,000	$0	$0	$1,720
TELEPHONE TIE-IN	1 LS	24	$30.00	$720	$1,500	$1,500	$0	$0	$2,220
FENCING	2,800 LF						$12	$33,600	$33,600
TOTAL, INITIAL SITEWORK				$4,944		$5,550		$35,400	$45,894
DECONTAMINATION PAD									
GRADE AREA	1 LS	24	$30.00	$720	$250	$250	$0	$0	$970
INSTALL HDPE LINER	1 LS	24	$30.00	$720	$1,000	$1,000	$0	$0	$1,720
GRAVEL	50 CY	24	$30.00	$720	$12	$600	$0	$0	$1,320
PIPING	1 LS	24	$30.00	$720	$1,500	$1,500	$0	$0	$2,220
SET UP TANKS	2 EA	24	$30.00	$720	$250	$500	$0	$0	$1,220
TOTAL, DECON. PAD				$3,600		$3,850		$0	$7,450
AIR MONITORING SYSTEM									
SET UP AIR MONITORS	2 EA	24	$30.00	$720	$100	$200	$150	$300	$1,220
SET UP MET. STATION	1 EA	24	$30.00	$720	$100	$100	$50	$50	$870
INITIAL SAMPLES	10 EA	1	$30.00	$30	$15	$150	$100	$1,000	$1,180
TOTAL, AIR MONITOR SYSTEM				$1,470		$450		$1,350	$3,270
TOTAL, MOBILIZATION				$12,414		$10,550		$40,050	$63,014

Appendix 5A: *(continued)*

DESCRIPTION	QUANTITY	HOURS	LABOR RATE	TOTAL LABOR	MATERIAL UNIT	TOTAL	SUBCONTRACTED UNIT	TOTAL	TOTAL
REMEDIATION									
EXCAVATION									
BULLDOZERS (3)	35 DAY	840	$32.00	$26,880	$0	$0	$750	$26,250	$53,130
FRONT END L'DERS (2)	35 DAY	560	$32.00	$17,920	$0	$0	$900	$31,500	$49,420
DUMP TRUCKS (6)	35 DAY	1680	$32.00	$53,760	$0	$0	$600	$21,000	$74,760
LABORERS	35 DAY	280	$32.00	$8,960	$0	$0	$0	$0	$8,960
				---------		-----------		------------	------------
TOTAL, EXCAVATION				$107,520		$0		$78,750	$186,270
TRANSPORTATION + DISPOSAL									
ROLL -OFF RENTAL	1,750 EA						$200	$350,000	$350,000
TRANSPORTATION	35,000 TN						$50	$1,750,000	$1,750,000
DISPOSAL	35,000 TN						$350	$12,250,000	$12,250,000
				---------		-----------		------------	------------
TOTAL, TRANSPORTATION + DISPOSAL				$0		$0		$14,350,000	$14,350,000
BACKFILL									
BULLDOZERS (3)	25 DAY	600	$30.00	$18,000	$0	$0	$750	$18,750	$36,750
ROLLER COMPACTOR (2)	25 DAY	400	$30.00	$12,000	$0	$0	$600	$15,000	$27,000
SOIL -DELIVERED	25,000 CY				$12	$300,000	$0	$0	$300,000
WATER TRUCK	25 DAY	200	$30.00	$6,000	$0	$0	$250	$6,250	$12,250
LABORERS	25 DAY	200	$30.00	$6,000	$0	$0	$0	$0	$6,000
				---------		-----------		------------	------------
TOTAL, BACKFILL				$42,000		$300,000		$40,000	$382,000
MISC ACTIVITES									
AIR SAMPLES (10/DAY)	80 DAY				$100	$8,000	$1,000	$80,000	$88,000
MISC DIPOSAL -PPE	50 DRUM				$30	$1,500	$250	$12,500	$14,000
COMPACTION TESTS	30 EA				$0	$0	$250	$7,500	$7,500
SOIL SAMPLING	100 EA				$10	$1,000	$150	$15,000	$16,000
				---------		-----------		------------	------------
TOTAL, MISC ACTIVITIES				$0		$10,500		$115,000	$125,500
				---------		-----------		------------	------------
TOTAL, REMEDIATION WORK				$149,520		$310,500		$14,583,750	$15,043,770

Appendix 5A: (*continued*)

DESCRIPTION	QUANTITY	HOURS	LABOR RATE	TOTAL LABOR	MATERIAL UNIT	TOTAL	SUBCONTRACTED UNIT	TOTAL	TOTAL
INDIRECT ACTIVITIES									
HEALTH + SAFETY SUPPLIES									
LEVEL "D" MOD OUTFITS	250 EA				$20	$5,000	$0	$0	$5,000
LEVEL "C" OUTFITS	500 EA				$35	$17,500	$0	$0	$17,500
AIR MONITORING EQUIP	6 MO				$5,000	$30,000	$0	$0	$30,000
DRUMS	100 EA				$30	$3,000	$0	$0	$3,000
SMALL TOOLS + CONSUMAB	1 LS				$10,000	$10,000	$0	$0	$10,000
TOTAL, HEALTH & SAFETY				$0		$65,500		$0	$65,500
TRAILERS + TEMP. FACILITIES									
OFFICE	6 MO				$250	$1,500	$0	$0	$1,500
DECONTAMINATION	6 MO				$900	$5,400	$0	$0	$5,400
EQUIP/TOOL	6 MO				$100	$600	$0	$0	$600
PORT-A-JOHNS	6 MO				$150	$900	$0	$0	$900
SIGNS	1 LS				$5,000	$5,000	$0	$0	$5,000
TOTAL, TRAILERS				$0		$13,400		$0	$13,400
UTILITIES									
ELECTRIC	6 MO				$500	$3,000	$0	$0	$3,000
WATER	6 MO				$200	$1,200	$0	$0	$1,200
TELEPHONE	6 MO				$500	$3,000	$0	$0	$3,000
RADIOS	1 LS				$2,500	$2,500	$0	$0	$2,500
FUEL	6 MO				$1,000	$6,000	$0	$0	$6,000
TOTAL, UTILITIES				$0		$15,700		$0	$15,700
MANAGEMENT PERSONNEL									
SITE MANAGER	1 LS	960	$40.00	$38,400	$10	$10	$0	$0	$38,410
SITE ENGINEER	1 LS	960	$30.00	$28,800	$6	$6	$0	$0	$28,806
SAFETY/HEALTH OFFICER	1 LS	960	$30.00	$28,800	$6	$6	$0	$0	$28,806
DECON. PERSON	1 LS	960	$30.00	$28,800	$0	$0	$0	$0	$28,800
SAMPLING TECHNICIAN	1 LS	960	$20.00	$19,200	$0	$0	$0	$0	$19,200
TOTAL, MANAGEMENT PERSONNEL				$144,000		$22		$0	$144,022
TOTAL, INDIRECTS				$144,000		$94,622		$0	$238,622

Appendix 5A: (*continued*)

DESCRIPTION	QUANTITY	HOURS	LABOR RATE	TOTAL LABOR	MATERIAL UNIT	MATERIAL TOTAL	SUBCONTRACTED UNIT	SUBCONTRACTED TOTAL	TOTAL
DEMOBILIZATION									
REMOVE TRAILERS									
PERSONNEL	1 EA	24	$30.00	$720	$0	$0	$250	$250	$970
DECONTAMINATION	1 EA	48	$30.00	$1,440	$0	$0	$250	$250	$1,690
EQUIP/TOOLS	1 EA	16	$30.00	$480	$0	$0	$150	$150	$630
PORT-A-JOHN	1 LS	0	$30.00	$0	$0	$0	$100	$100	$100
TOTAL, TRAILERS				$2,640		$0		$750	$3,390
MOVE OUT EQUIPMENT									
DECON. EQUIPMENT	14 EA	64	$30.00	$1,920	$0	$0	$0	$0	$1,920
HAUL OUT	7 EA	0	$30.00	$0	$0	$0	$250	$1,750	$1,750
TRUCKS	7 EA	0	$30.00	$0	$0	$0	$100	$700	$700
TOTAL, EQUIPMENT				$1,920		$0		$2,450	$4,370
REMOVE DECON. FACILITIES									
REMOVE PAD	1 LS	48	$30.00	$500	$500	$500	$100	$100	$1,100
REMOVE PIPING	1 LS	60	$30.00	$250	$0	$0	$100	$100	$350
DISPOSE OF WATER	10,000 GA	24	$30.00	$720	$0	$0	$3	$30,000	$30,720
REMOVE TANKS	1 LS	24	$30.00	$720	$0	$0	$0	$0	$720
TOTAL, DECON. FACILITIES				$2,190		$500		$30,200	$32,890
TOTAL, DEMOBILIZATION				$6,750		$500		$33,400	$40,650

Appendix 5B: Example of Conceptual Estimate Installation of a Landfill Capping System

DESCRIPTION	QUANTITY	HOURS	LABOR RATE	TOTAL LABOR	MATERIAL UNIT	TOTAL	SUBCONTRACTED UNIT	TOTAL	TOTAL
SUMMARY									
TOTAL, PREMOBILIZATION ACTIVITIES				$55,160		$523,650		$0	$578,810
TOTAL, GENERAL MOBILIZATION				$12,158		$10,550		$39,400	$62,108
TOTAL, INCINERATOR SYSTEM MOB.				$1,182,000		$591,300		$1,682,600	$3,455,900
TOTAL, REMEDIATION WORK				$2,494,440		$1,733,500		$4,786,500	$9,014,440
TOTAL, INDIRECTS				$360,000		$346,522		$0	$706,522
TOTAL, DEMOBILIZATION				$155,790		$50,500		$337,750	$544,040
				-----------		-----------		-----------	-----------
SUBTOTAL				$4,259,548		$3,256,022		$6,846,250	$14,361,820
CONTINGENCY	20.00%								$2,872,364
FEE	10.00%								$1,436,182
				-----------		-----------		-----------	-----------
TOTAL, ESTIMATE				$4,259,548		$3,256,022		$6,846,250	$18,670,366

PRE-MOBILIZATION ACTIVITIES

DESCRIPTION	QUANTITY	HOURS	LABOR RATE	TOTAL LABOR	MATERIAL UNIT	TOTAL	SUBCONTRACTED	TOTAL
WRITING OF PLANS								
SITE HEALTH & SAFETY	1 LS	80	$40.00	$3,200				$3,200
WORK PLAN	1 LS	80	$40.00	$3,200				$3,200
QUALITY ASSURANCE	1 LS	60	$40.00	$2,400				$2,400
AIR MONITORING	1 LS	60	$40.00	$2,400				$2,400
SITE SECURITY	1 LS	20	$40.00	$800				$800
OBTAIN LOCAL PERMITS	1 LS	80	$40.00	$3,200				$3,200
TEST BURN REPORT	1 LS	640	$40.00	$25,600				$25,600
COPYING/PUBLICATIONS	1 LS				$10,000	$10,000		$10,000
				-----------		-----------		-----------
TOTAL, WRITING OF PLANS				$40,800		$10,000	$0	$50,800
HOME OFFICE SERVICES								
PROCUREMENT	1 LS	80	$30.00	$2,400				$2,400
PROJECT CONTROLS	1 LS	40	$30.00	$1,200				$1,200
GENERAL MANAGEMENT	1 LS	20	$50.00	$1,000				$1,000
				-----------		-----------		-----------
TOTAL, HOME OFFICE SERVICES				$4,600		$0	$0	$4,600
CFR 1910.120 MEDICAL/TRAINING								
MEDICAL EXAMS	12 EA	48	$20.00	$960	$500	$6,000		$6,960
40 HR TRAINING	4 EA	160	$20.00	$3,200	$600	$2,400		$5,600
8 HR REFRESHER	5 EA	40	$20.00	$800	$250	$1,250		$2,050
				-----------		-----------		-----------
TOTAL, CFR 1910.120				$4,960		$9,650	$0	$14,610
PRECONSTRUCTION MEETINGS								
PROJECT MANAGER	1 LS	32	$50.00	$1,600				$1,600
PROJECT ENGINEER	1 LS	48	$40.00	$1,920				$1,920
SITE MANAGER	1 LS	32	$40.00	$1,280				$1,280
ALLOW FOR TRAVEL	1 LS				$4,000	$4,000		$4,000
				-----------		-----------		-----------
TOTAL, PRECONSTRUCTION MEETINGS				$4,800		$4,000	$0	$8,800
PERFORMANCE & PAYMENT BOND								
ALLOWANCE FOR	1 LS					$500,000		$500,000
				-----------		-----------		-----------
TOTAL, PREMOBILIZATION ACTIVITIES				$55,160		$523,650	$0	$578,810

Appendix 5B: (*continued*)

DESCRIPTION	QUANTITY	HOURS	LABOR RATE	TOTAL LABOR	MATERIAL UNIT	MATERIAL TOTAL	SUBCONTRACTED UNIT	SUBCONTRACTED TOTAL	TOTAL
GENERAL MOBILIZATION ACTIVITIES									
MOVE IN + SET UP TRAILERS									
OFFICE	1 EA	24	$30.00	$720	$250	$250	$250	$250	$1,220
DECONTAMINATION	1 EA	48	$30.00	$1,440	$300	$300	$500	$500	$2,240
TOOL/EQUIPMENT	1 EA	8	$30.00	$240	$150	$150	$100	$100	$490
TOTAL, MOVE IN TRAILERS				$2,400		$700		$850	$3,950
MOVE IN EQUIPMENT									
BULLDOZERS	3 EA						$250	$750	$750
DUMP & WATER TRUCKS	3 EA						$100	$300	$300
FRONT END LOADERS	1 EA						$250	$250	$250
ROLLER COMPACTOR	2 EA						$250	$500	$500
TOTAL, MOVE IN EQUIPMENT				$0		$0		$1,800	$1,800
SET UP WORK ZONES + INITIAL SITEWORK									
SURVEYOR	2 DAY						$650	$1,300	$1,300
TAPES AND STAKES	1 LS				$150	$150			$150
CLEARING + GRUBBING	1 LS	32	$32.00	$1,024	$250	$250	$250	$250	$1,524
GRADING	1 LS	32	$32.00	$1,024	$250	$250	$250	$250	$1,524
GRAVEL PAVING	200 CY	40	$30.00	$1,200	$12	$2,400	$0	$0	$3,600
ELECTRIC CONNECTIONS	1 LS	24	$30.00	$720	$1,000	$1,000	$0	$0	$1,720
TELEPHONE TIE-IN	1 LS	24	$30.00	$720	$1,500	$1,500	$0	$0	$2,220
FENCING	2,800 LF						$12	$33,600	$33,600
TOTAL, INITIAL SITEWORK				$4,688		$5,550		$35,400	$45,638
DECONTAMINATION PAD									
GRADE AREA	1 LS	24	$30.00	$720	$250	$250	$0	$0	$970
INSTALL HDPE LINER	1 LS	24	$30.00	$720	$1,000	$1,000	$0	$0	$1,720
GRAVEL	50 CY	24	$30.00	$720	$12	$600	$0	$0	$1,320
PIPING	1 LS	24	$30.00	$720	$1,500	$1,500	$0	$0	$2,220
SET UP TANKS	2 EA	24	$30.00	$720	$250	$500	$0	$0	$1,220
TOTAL, DECON. PAD				$3,600		$3,850		$0	$7,450
AIR MONITORING SYSTEM									
SET UP AIR MONITORS	2 EA	24	$30.00	$720	$100	$200	$150	$300	$1,220
SET UP MET. STATION	1 EA	24	$30.00	$720	$100	$100	$50	$50	$870
INITIAL SAMPLES	10 EA	1	$30.00	$30	$15	$150	$100	$1,000	$1,180
TOTAL, AIR MONITOR SYSTEM				$1,470		$450		$1,350	$3,270
TOTAL, GENERAL MOBILIZATION				$12,158		$10,550		$39,400	$62,108

Appendix 5B: (continued)

DESCRIPTION	QUANTITY	HOURS	LABOR RATE	TOTAL LABOR	MATERIAL UNIT	TOTAL	SUBCONTRACTED UNIT	TOTAL	TOTAL
INCINERATOR SYSTEM MOBILIZATION									
MOVE INCINERATOR ONSITE & SET UP									
FREIGHT -TRUCKS	1 LS					$6,000	$40,000	$40,000	$46,000
MECHANICAL WORK	1 LS	960	$30.00	$28,800	$35,000	$35,000	$10,000	$10,000	$73,800
ELECTRICAL WORK	1 LS	640	$30.00	$19,200	$25,000	$25,000	$5,000	$5,000	$49,200
CONCRETE FOUNDATIONS	100 CY	600	$30.00	$18,000	$110	$11,000	$0	$0	$29,000
REFRACTORY/MISC REPAIR	1 LS						$50,000	$50,000	$50,000
TOTAL, INCINERATOR ONSITE				$66,000		$77,000		$105,000	$248,000
INSTALL SUPPORTING SYSTEMS									
WATER TREATMENT SYSTEM	1 LS	480	$30.00	$14,400	$30,000	$30,000	$0	$0	$44,400
CEM TRAILER	1 LS	48	$30.00	$1,440	$10,000	$10,000	$0	$0	$11,440
POLLUTION CONTROL SYS.	1 LS	240	$30.00	$7,200	$15,000	$15,000	$0	$0	$22,200
ASH HANDLING SYSTEM	1 LS	240	$30.00	$7,200	$25,000	$25,000	$0	$0	$32,200
MATERIAL FEED SYSTEM	1 LS	120	$30.00	$3,600	$10,000	$10,000	$0	$0	$13,600
TOTAL, SUPPORTING SYSTEM				$33,840		$90,000		$0	$123,840
SHAKEDOWN/START UP THE UNIT									
PERSONNEL	49 DAY	12,250	$30.00	$367,500	$0	$0	$0	$0	$367,500
UTILITIES	49 DAY						$1,500	$73,500	$73,500
MISC EXPENSES	49 DAY				$1,200	$58,800	$500	$24,500	$83,300
CHEMICALS	49 DAY				$250	$12,250	$100	$4,900	$17,150
LABORATORY	49 DAY				$150	$7,350	$3,000	$147,000	$154,350
DEPRECIATION	49 DAY						$3,000	$147,000	$147,000
MAINTENANCE	49 DAY						$1,000	$49,000	$49,000
TOTAL, SHAKEDOWN/START UP				$367,500		$78,400		$445,900	$891,800
TRIAL BURN									
PERSONNEL -WORKERS	7 DAY	1,750	$30.00	$52,500	$0	$0	$0	$0	$52,500
PERSONNEL - TESTING	7 DAY	700	$30.00	$21,000	$1,300	$9,100	$0	$0	$30,100
UTILITIES	7 DAY						$1,500	$10,500	$10,500
MISC EXPENSES	7 DAY				$3,000	$21,000	$500	$3,500	$24,500
CHEMICALS	7 DAY				$250	$1,750	$100	$700	$2,450
LABORATORY	7 DAY				$150	$1,050	$5,000	$35,000	$36,050
DEPRECIATION	7 DAY						$3,000	$21,000	$21,000
MAINTENANCE	7 DAY						$1,000	$7,000	$7,000
SPECIAL TEST/ANALYTICS	1 LS					$100,000		$250,000	$350,000
TOTAL, TRIAL BURN				$73,500		$132,900		$327,700	$534,100

Appendix 5B: (continued)

DESCRIPTION	QUANTITY	HOURS	LABOR RATE	TOTAL LABOR	MATERIAL UNIT	MATERIAL TOTAL	SUBCONTRACTED UNIT	SUBCONTRACTED TOTAL	TOTAL
STAND BY (HOT)									
PERSONNEL -WORKERS	30 DAY	7,500	$30.00	$225,000	$0	$0	$0	$0	$225,000
UTILITIES	30 DAY						$1,500	$45,000	$45,000
MISC EXPENSES	30 DAY				$3,000	$90,000	$500	$15,000	$105,000
CHEMICALS	30 DAY				$250	$7,500	$100	$3,000	$10,500
LABORATORY	30 DAY				$50	$1,500	$500	$15,000	$16,500
DEPRECIATION	30 DAY						$3,000	$90,000	$90,000
MAINTENANCE	30 DAY						$1,000	$30,000	$30,000
TOTAL, STAND BY				$225,000		$99,000		$198,000	$522,000
INTERIM OPERATIONS - 50% OF CAPACITY									
PERSONNEL -WORKERS	60 DAY	15,000	$30.00	$450,000	$0	$0	$0	$0	$450,000
UTILITIES	60 DAY						$1,000	$60,000	$60,000
MISC EXPENSES	60 DAY				$3,000	$180,000	$500	$30,000	$210,000
CHEMICALS	60 DAY				$250	$15,000	$100	$6,000	$21,000
LABORATORY	60 DAY				$150	$9,000	$4,500	$270,000	$279,000
DEPRECIATION	60 DAY						$3,000	$180,000	$180,000
MAINTENANCE	60 DAY						$1,000	$60,000	$60,000
TOTAL, INTERIM OPERATIONS				$450,000		$204,000		$606,000	$1,260,000
TOTAL, INCINERATOR SYSTEM MOB.				$1,182,000		$591,300		$1,682,600	$3,455,900

Appendix 5B: (*continued*)

DESCRIPTION	QUANTITY	HOURS	LABOR RATE	TOTAL LABOR	MATERIAL UNIT	MATERIAL TOTAL	SUBCONTRACTED UNIT	SUBCONTRACTED TOTAL	TOTAL
REMEDIATION									
EXCAVATION									
BULLDOZERS (1)	300 DAY	2400	$32.00	$76,800	$0	$0	$750	$225,000	$301,800
FRONT END L'DERS (1)	300 DAY	2400	$32.00	$76,800	$0	$0	$900	$270,000	$346,800
DUMP TRUCKS (2)	300 DAY	4800	$32.00	$153,600	$0	$0	$600	$180,000	$333,600
LABORERS	300 DAY	2400	$32.00	$76,800	$0	$0	$0	$0	$76,800
TOTAL, EXCAVATION				$384,000		$0		$675,000	$1,059,000
OPERATIONS OF INCINERATOR									
PERSONNEL -WORKERS	270 DAY	67,500	$30.00	$2,025,000	$0	$0	$0	$0	$2,025,000
UTILITIES	270 DAY						$1,800	$486,000	$486,000
MISC EXPENSES	270 DAY				$5,000	$1,350,000	$500	$135,000	$1,485,000
CHEMICALS	270 DAY				$600	$162,000	$100	$27,000	$189,000
LABORATORY	270 DAY				$200	$54,000	$6,000	$1,620,000	$1,674,000
DEPRECIATION	270 DAY						$3,000	$810,000	$810,000
MAINTENANCE	270 DAY						$1,000	$270,000	$270,000
TOTAL, INCINERATOR OPERATIONS				$2,025,000		$1,566,000		$3,348,000	$6,939,000
BACKFILL - ASH & FINAL COVER									
BULLDOZERS (3)	45 DAY	1080	$32.00	$34,560	$0	$0	$750	$33,750	$68,310
ROLLER COMPACTOR (2)	45 DAY	720	$32.00	$23,040	$0	$0	$600	$27,000	$50,040
SOIL -DELIVERED	6,000 CY				$12	$72,000	$0	$0	$72,000
WATER TRUCK	45 DAY	360	$32.00	$11,520	$0	$0	$250	$11,250	$22,770
LABORERS	45 DAY	360	$32.00	$11,520	$0	$0	$0	$0	$11,520
TOTAL, BACKFILL				$80,640		$72,000		$72,000	$224,640
LINING OPERATION									
MEMBRANE LINER	170,000 SF						$1	$127,500	$127,500
SOIL BASE	6,000 CY	160	$30.00	$4,800	$12	$72,000			$76,800
GEONET/FABRIC FILTER	170,000 SF						$2.50	$425,000	$425,000
TOTAL, LINING OPERATION				$4,800		$72,000		$552,500	$629,300
MISC ACTIVITES									
AIR SAMPLES (2/DAY)	320 DAY				$50	$16,000	$200	$64,000	$80,000
COMPACTION TEST	50 EA				$0	$0	$250	$12,500	$12,500
MISC DISPOSAL -PPE	250 DRUM				$30	$7,500	$250	$62,500	$70,000
				$0		$23,500		$139,000	$162,500
TOTAL, REMEDIATION WORK				$2,494,440		$1,733,500		$4,786,500	$9,014,440

Appendix 5B: *(continued)*

DESCRIPTION	QUANTITY	HOURS	LABOR RATE	TOTAL LABOR	MATERIAL UNIT	TOTAL	SUBCONTRACTED UNIT	TOTAL	TOTAL
INDIRECT ACTIVITIES									
HEALTH + SAFETY SUPPLIES									
LEVEL "D" MOD OUTFITS	7,000 EA				$20	$140,000	$0	$0	$140,000
LEVEL "C" OUTFITS	1,500 EA				$35	$52,500	$0	$0	$52,500
AIR MONITORING EQUIP	12 MO				$5,000	$60,000	$0	$0	$60,000
DRUMS	250 EA				$30	$7,500	$0	$0	$7,500
SMALL TOOLS + CONSUMAB	1 LS				$25,000	$25,000	$0	$0	$25,000
TOTAL, HEALTH & SAFETY				$0		$285,000		$0	$285,000
TRAILERS + TEMP. FACILITIES									
OFFICE	15 MO				$250	$3,750	$0	$0	$3,750
DECONTAMINATION	15 MO				$900	$13,500	$0	$0	$13,500
EQUIP/TOOL	15 MO				$100	$1,500	$0	$0	$1,500
PORT-A-JOHNS	15 MO				$150	$2,250	$0	$0	$2,250
SIGNS	1 LS				$5,000	$5,000	$0	$0	$5,000
TOTAL, TRAILERS				$0		$26,000		$0	$26,000
UTILITIES - TRAILERS ONLY									
ELECTRIC	15 MO				$500	$7,500	$0	$0	$7,500
WATER	15 MO				$200	$3,000	$0	$0	$3,000
TELEPHONE	15 MO				$500	$7,500	$0	$0	$7,500
RADIOS	1 LS				$2,500	$2,500	$0	$0	$2,500
FUEL	15 MO				$1,000	$15,000	$0	$0	$15,000
TOTAL, UTILITIES				$0		$35,500		$0	$35,500
MANAGEMENT PERSONNEL									
SITE MANAGER	1 LS	2400	$40.00	$96,000	$10	$10	$0	$0	$96,010
SITE ENGINEER	1 LS	2400	$30.00	$72,000	$6	$6	$0	$0	$72,006
SAFETY/HEALTH OFFICER	1 LS	2400	$30.00	$72,000	$6	$6	$0	$0	$72,006
DECON. PERSON	1 LS	2400	$30.00	$72,000	$0	$0	$0	$0	$72,000
SAMPLING TECHNICIAN	1 LS	2400	$20.00	$48,000	$0	$0	$0	$0	$48,000
TOTAL, MANAGEMENT PERSONNEL				$360,000		$22		$0	$360,022
TOTAL, INDIRECTS				$360,000		$346,522		$0	$706,522

Appendix 5B: *(continued)*

DESCRIPTION	QUANTITY	HOURS	LABOR RATE	TOTAL LABOR	MATERIAL UNIT	MATERIAL TOTAL	SUBCONTRACTED UNIT	SUBCONTRACTED TOTAL	TOTAL
DEMOBILIZATION									
REMOVE TRAILERS									
PERSONNEL	1 EA	24	$30.00	$720	$0	$0	$250	$250	$970
DECONTAMINATION	1 EA	48	$30.00	$1,440	$0	$0	$250	$250	$1,690
EQUIP/TOOLS	1 EA	16	$30.00	$480	$0	$0	$150	$150	$630
PORT-A-JOHN	1 LS	0	$30.00	$0	$0	$0	$100	$100	$100
TOTAL, TRAILERS				$2,640		$0		$750	$3,390
MOVE OUT EQUIPMENT									
DECON. EQUIPMENT	9 EA	40	$30.00	$1,200	$0	$0	$0	$0	$1,200
HAUL OUT EQUIPMENT	6 EA	0	$30.00	$0	$0	$0	$250	$1,500	$1,500
TRUCKS	3 EA	0	$30.00	$0	$0	$0	$100	$300	$300
TOTAL, EQUIPMENT				$1,200		$0		$1,800	$3,000
REMOVE DECON. FACILITIES									
REMOVE PAD	1 LS	48	$30.00	$500	$500	$500	$100	$100	$1,100
REMOVE PIPING	1 LS	60	$30.00	$250	$0	$0	$100	$100	$350
DISPOSE OF WATER	10,000 GA	24	$20.00	$480	$0	$0	$3	$30,000	$30,480
REMOVE TANKS	1 LS	24	$30.00	$720	$0	$0	$0	$0	$720
TOTAL, DECON. FACILITIES				$1,950		$500		$30,200	$32,650
REMOVE INCINERATOR									
PERSONNEL -WORKERS	50 DAY	5,000	$30.00	$150,000	$0	$0	$0	$0	$150,000
MECHANICAL SUBCONTRACT	1 LS						$50,000	$50,000	$50,000
MISC EXPENSES	50 DAY				$1,000	$50,000	$500	$25,000	$75,000
ELECTRIC SUBCONTRACTOR	1 LS						$30,000	$30,000	$30,000
DEPRECIATION	50 DAY						$3,000	$150,000	$150,000
MAINTENANCE	50 DAY						$1,000	$50,000	$50,000
TOTAL, INCINERATOR REMOVAL				$150,000		$50,000		$305,000	$505,000
TOTAL, DEMOBILIZATION				$155,790		$50,500		$337,750	$544,040

Appendix 5C: Example of Conceptual Estimate Onsite Incineration of
Contaminates

DESCRIPTION	QUANTITY	HOURS	LABOR RATE	TOTAL LABOR	MATERIAL UNIT	TOTAL	SUBCONTRACTED UNIT	TOTAL	TOTAL
SUMMARY									
TOTAL, PREMOBILIZATION ACTIVITIES				$29,560		$115,650		$0	$145,210
TOTAL, MOBILIZATION				$11,646		$10,550		$39,650	$61,846
TOTAL, REMEDIATION WORK				$68,800		$253,500		$1,645,700	$1,968,000
TOTAL, INDIRECTS				$134,400		$67,122		$0	$201,522
TOTAL, DEMOBILIZATION				$6,030		$500		$18,000	$24,530
SUBTOTAL				$250,436		$447,322		$1,703,350	$2,401,108
CONTINGENCY	20.00%								$480,222
FEE	10.00%								$240,111
TOTAL, ESTIMATE				$250,436		$447,322		$1,703,350	$3,121,440

DESCRIPTION	QUANTITY	HOURS	LABOR RATE	TOTAL LABOR	MATERIAL UNIT	TOTAL	SUBCONTRACTED UNIT	TOTAL	TOTAL
PRE-MOBILIZATION ACTIVITIES									
WRITING OF PLANS									
SITE HEALTH & SAFETY	1 LS	80	$40.00	$3,200					$3,200
WORK PLAN	1 LS	80	$40.00	$3,200					$3,200
QUALITY ASSURANCE	1 LS	60	$40.00	$2,400					$2,400
AIR MONITORING	1 LS	60	$40.00	$2,400					$2,400
SITE SECURITY	1 LS	20	$40.00	$800					$800
OBTAIN LOCAL PERMITS	1 LS	80	$40.00	$3,200					$3,200
COPYING/PUBLICATIONS	1 LS				$2,000	$2,000			$2,000
TOTAL, WRITING OF PLANS				$15,200		$2,000		$0	$17,200
HOME OFFICE SERVICES									
PROCUREMENT	1 LS	80	$30.00	$2,400					$2,400
PROJECT CONTROLS	1 LS	40	$30.00	$1,200					$1,200
GENERAL MANAGEMENT	1 LS	20	$50.00	$1,000					$1,000
TOTAL, HOME OFFICE SERVICES				$4,600		$0		$0	$4,600
CFR 1910.120 MEDICAL/TRAINING									
MEDICAL EXAMS	12 EA	48	$20.00	$960	$500	$6,000			$6,960
40 HR TRAINING	4 EA	160	$20.00	$3,200	$600	$2,400			$5,600
8 HR REFRESHER	5 EA	40	$20.00	$800	$250	$1,250			$2,050
TOTAL, CFR 1910.120				$4,960		$9,650		$0	$14,610
PRECONSTRUCTION MEETINGS									
PROJECT MANAGER	1 LS	32	$50.00	$1,600					$1,600
PROJECT ENGINEER	1 LS	48	$40.00	$1,920					$1,920
SITE MANAGER	1 LS	32	$40.00	$1,280					$1,280
ALLOW FOR TRAVEL	1 LS				$4,000	$4,000			$4,000
TOTAL, PRECONSTRUCTION MEETINGS				$4,800		$4,000		$0	$8,800
PERFORMANCE & PAYMENT BOND									
ALLOWANCE FOR	1 LS					$100,000			$100,000
TOTAL, PREMOBILIZATION ACTIVITIES				$29,560		$115,650		$0	$145,210

Appendix 5C: (*continued*)

DESCRIPTION	QUANTITY	HOURS	LABOR RATE	TOTAL LABOR	MATERIAL UNIT	TOTAL	SUBCONTRACTED UNIT	TOTAL	TOTAL
GENERAL MOBILIZATION ACTIVITIES									
MOVE IN + SET UP TRAILERS									
OFFICE	1 EA	24	$30.00	$720	$250	$250	$250	$250	$1,220
DECONTAMINATION	1 EA	48	$30.00	$1,440	$300	$300	$500	$500	$2,240
TOOL/EQUIPMENT	1 EA	8	$30.00	$240	$150	$150	$100	$100	$490
TOTAL, MOVE IN TRAILERS				$2,400		$700		$850	$3,950
MOVE IN EQUIPMENT									
BULLDOZERS	3 EA						$250	$750	$750
DUMP + WATER TRUCKS	3 EA						$100	$300	$300
FRONT END LOADERS	2 EA						$250	$500	$500
ROLLER COMPACTORS	2 EA						$250	$500	$500
TOTAL, MOVE IN EQUIPMENT				$0		$0		$2,050	$2,050
SET UP WORK ZONES + INITIAL SITEWORK									
SURVEYOR	2 DAY						$650	$1,300	$1,300
TAPES AND STAKES	1 LS				$150	$150			$150
CLEARING + GRUBBING	1 LS	24	$32.00	$768	$250	$250	$250	$250	$1,268
GRADING	1 LS	24	$32.00	$768	$250	$250	$250	$250	$1,268
GRAVEL PAVING	200 CY	40	$30.00	$1,200	$12	$2,400	$0	$0	$3,600
ELECTRIC CONNECTIONS	1 LS	24	$30.00	$720	$1,000	$1,000	$0	$0	$1,720
TELEPHONE TIE-IN	1 LS	24	$30.00	$720	$1,500	$1,500	$0	$0	$2,220
FENCING	2,800 LF						$12	$33,600	$33,600
TOTAL, INITIAL SITEWORK				$4,176		$5,550		$35,400	$45,126
DECONTAMINATION PAD									
GRADE AREA	1 LS	24	$30.00	$720	$250	$250	$0	$0	$970
INSTALL HDPE LINER	1 LS	24	$30.00	$720	$1,000	$1,000	$0	$0	$1,720
GRAVEL	50 CY	24	$30.00	$720	$12	$600	$0	$0	$1,320
PIPING	1 LS	24	$30.00	$720	$1,500	$1,500	$0	$0	$2,220
SET UP TANKS	2 EA	24	$30.00	$720	$250	$500	$0	$0	$1,220
TOTAL, DECON. PAD				$3,600		$3,850		$0	$7,450
AIR MONITORING SYSTEM									
SET UP AIR MONITORS	2 EA	24	$30.00	$720	$100	$200	$150	$300	$1,220
SET UP MET. STATION	1 EA	24	$30.00	$720	$100	$100	$50	$50	$870
INITIAL SAMPLES	10 EA	1	$30.00	$30	$15	$150	$100	$1,000	$1,180
TOTAL, AIR MONITOR SYSTEM				$1,470		$450		$1,350	$3,270
TOTAL, MOBILIZATION				$11,646		$10,550		$39,650	$61,846

Appendix 5C: *(continued)*

DESCRIPTION	QUANTITY	HOURS	LABOR RATE	TOTAL LABOR	MATERIAL UNIT	TOTAL	SUBCONTRACTED UNIT	TOTAL	TOTAL
REMEDIATION									
EXCAVATION									
BULLDOZERS (3)	10 DAY	240	$32.00	$7,680	$0	$0	$750	$7,500	$15,180
FRONT END L'DERS (2)	10 DAY	160	$32.00	$5,120	$0	$0	$900	$9,000	$14,120
DUMP TRUCKS (6)	10 DAY	480	$32.00	$15,360	$0	$0	$600	$6,000	$21,360
LABORERS	10 DAY	80	$32.00	$2,560	$0	$0	$0	$0	$2,560
TOTAL, EXCAVATION				$30,720		$0		$22,500	$53,220
BACKFILL - INITIAL COVER -12"									
BULLDOZERS (3)	10 DAY	240	$32.00	$7,680	$0	$0	$750	$7,500	$15,180
ROLLER COMPACTOR (2)	10 DAY	160	$32.00	$5,120	$0	$0	$600	$6,000	$11,120
SOIL -DELIVERED	6,500 CY				$12	$78,000	$0	$0	$78,000
WATER TRUCK	10 DAY	80	$32.00	$2,560	$0	$0	$250	$2,500	$5,060
LABORERS	10 DAY	80	$32.00	$2,560	$0	$0	$0	$0	$2,560
TOTAL, BACKFILL				$17,920		$78,000		$16,000	$111,920
LINING SYSTEM									
GEO-NET	180,000 SF						$1.00	$180,000	$180,000
FABRIC FILTER	180,000 SF						$1.50	$270,000	$270,000
MEMBRANE	180,000 SF						$0.75	$135,000	$135,000
TOTAL, LINING SYSTEM				$0		$0		$585,000	$585,000
BACKFILL - FINAL COVER - 24"									
BULLDOZERS (3)	12 DAY	288	$30.00	$8,640	$0	$0	$750	$9,000	$17,640
ROLLER COMPACTOR (2)	12 DAY	192	$30.00	$5,760	$0	$0	$600	$7,200	$12,960
SOIL -DELIVERED	13,000 CY				$12	$156,000	$0	$0	$156,000
WATER TRUCK	12 DAY	96	$30.00	$2,880	$0	$0	$250	$3,000	$5,880
LABORERS	12 DAY	96	$30.00	$2,880	$0	$0	$0	$0	$2,880
SEEDING	180,000 SF						$0	$18,000	$18,000
TOTAL, BACKFILL				$20,160		$156,000		$37,200	$213,360
MISC ACTIVITES									
AIR SAMPLES (10/DAY)	60 DAY				$100	$6,000	$1,000	$60,000	$66,000
MONITORING WELLS	120 EA				$100	$12,000	$7,500	$900,000	$912,000
COMPACTION TEST	50 EA				$0	$0	$250	$12,500	$12,500
MISC DISPOSAL -PPE	50 DRUM				$30	$1,500	$250	$12,500	$14,000
				$0		$19,500		$985,000	$1,004,500
TOTAL, REMEDIATION WORK				$68,800		$253,500		$1,645,700	$1,968,000

Appendix 5C: *(continued)*

DESCRIPTION	QUANTITY	HOURS	LABOR RATE	TOTAL LABOR	MATERIAL UNIT	MATERIAL TOTAL	SUBCONTRACTED UNIT	SUBCONTRACTED TOTAL	TOTAL
INDIRECT ACTIVITIES									
HEALTH + SAFETY SUPPLIES									
LEVEL "D" MOD OUTFITS	150 EA				$20	$3,000	$0	$0	$3,000
LEVEL "C" OUTFITS	100 EA				$35	$3,500	$0	$0	$3,500
AIR MONITORING EQUIP	4 MO				$5,000	$20,000	$0	$0	$20,000
DRUMS	50 EA				$30	$1,500	$0	$0	$1,500
SMALL TOOLS + CONSUMAB	1 LS				$10,000	$10,000	$0	$0	$10,000
TOTAL, HEALTH & SAFETY				$0		$38,000		$0	$38,000
TRAILERS + TEMP. FACILITIES									
OFFICE	6 MO				$250	$1,500	$0	$0	$1,500
DECONTAMINATION	6 MO				$900	$5,400	$0	$0	$5,400
EQUIP/TOOL	6 MO				$100	$600	$0	$0	$600
PORT-A-JOHNS	6 MO				$150	$900	$0	$0	$900
SIGNS	1 LS				$5,000	$5,000	$0	$0	$5,000
TOTAL, TRAILERS				$0		$13,400		$0	$13,400
UTILITIES									
ELECTRIC	6 MO				$500	$3,000	$0	$0	$3,000
WATER	6 MO				$200	$1,200	$0	$0	$1,200
TELEPHONE	6 MO				$500	$3,000	$0	$0	$3,000
RADIOS	1 LS				$2,500	$2,500	$0	$0	$2,500
FUEL	6 MO				$1,000	$6,000	$0	$0	$6,000
TOTAL, UTILITIES				$0		$15,700		$0	$15,700
MANAGEMENT PERSONNEL									
SITE MANAGER	1 LS	960	$40.00	$38,400	$10	$10	$0	$0	$38,410
SITE ENGINEER	1 LS	960	$30.00	$28,800	$6	$6	$0	$0	$28,806
SAFETY/HEALTH OFFICER	1 LS	960	$30.00	$28,800	$6	$6	$0	$0	$28,806
DECON. PERSON	1 LS	960	$20.00	$19,200	$0	$0	$0	$0	$19,200
SAMPLING TECHNICIAN	1 LS	960	$20.00	$19,200	$0	$0	$0	$0	$19,200
TOTAL, MANAGEMENT PERSONNEL				$134,400		$22		$0	$134,422
TOTAL, INDIRECTS				$134,400		$67,122		$0	$201,522

Appendix 5C: *(continued)*

DESCRIPTION	QUANTITY	HOURS	LABOR RATE	TOTAL LABOR	MATERIAL UNIT	TOTAL	SUBCONTRACTED UNIT	TOTAL	TOTAL
DEMOBILIZATION									
REMOVE TRAILERS									
PERSONNEL	1 EA	24	$30.00	$720	$0	$0	$250	$250	$970
DECONTAMINATION	1 EA	48	$30.00	$1,440	$0	$0	$250	$250	$1,690
EQUIP/TOOLS	1 EA	16	$30.00	$480	$0	$0	$150	$150	$630
PORT-A-JOHN	1 LS	0	$30.00	$0	$0	$0	$100	$100	$100
				---------		-----------		-----------	-----------
TOTAL, TRAILERS				$2,640		$0		$750	$3,390
MOVE OUT EQUIPMENT									
DECON. EQUIPMENT	10 EA	40	$30.00	$1,200	$0	$0	$0	$0	$1,200
HAUL OUT	7 EA	0	$30.00	$0	$0	$0	$250	$1,750	$1,750
TRUCKS	3 EA	0	$30.00	$0	$0	$0	$100	$300	$300
				---------		-----------		-----------	-----------
TOTAL, EQUIPMENT				$1,200		$0		$2,050	$3,250
REMOVE DECON. FACILITIES									
REMOVE PAD	1 LS	48	$30.00	$500	$500	$500	$100	$100	$1,100
REMOVE PIPING	1 LS	60	$30.00	$250	$0	$0	$100	$100	$350
DISPOSE OF WATER	5,000 GA	24	$30.00	$720	$0	$0	$3	$15,000	$15,720
REMOVE TANKS	1 LS	24	$30.00	$720	$0	$0	$0	$0	$720
				---------		-----------		-----------	-----------
TOTAL, DECON. FACILITIES				$2,190		$500		$15,200	$17,890
				---------		-----------		-----------	-----------
TOTAL, DEMOBILIZATION				$6,030		$500		$18,000	$24,530

6
Cost Drivers for Hazardous Waste Projects

Donald J. Cass
Cass & Associates, Los Angeles, California

6.1 INTRODUCTION

Current cost literature (at the time this chapter was prepared) does not provide a structured approach or methodology to evaluate documents, assumptions used, or basis of estimate for the resulting cost estimate prepared for a hazardous waste project. This chapter will share with environmental, estimating, cost engineering, and other industry professionals a structured approach to develop and evaluate unique cost drivers and other components of a hazardous waste site estimate. A structured approach provides a detailed audit trail for labor multipliers used in assessing labor productivity, as well as defining other cost drivers (personal protective equipment ensemble costs, decontamination facilities, and other required infrastructure, contractor's plant, etc.), which, when costed out, constitute an estimate for hazardous waste projects.

The following strategy is recommended for accomplishing the aforementioned. Section 6.2 outlines labor delays that are unique to all hazardous waste projects and a technique for evaluation. Section 6.3 discusses a method for calculating labor productivity multipliers encompassing worker protection Levels A, B, C, and D. Section 6.4 defines other cost drivers unique to hazardous waste projects. Section 6.5 addresses utilization of electronic media and Section 6.6 demonstrates an example utilizing labor multipliers developed for this chapter.

6.1.1 Elements of a Hazardous Waste Project Estimate

The following are elements of a hazardous waste project estimate (this is discussed further in Chapters 5, 7, and 8).

- Direct costs (labor, material, and equipment) for the approved remedial corrective action for the specific hazardous waste site project*
- Other applicable "direct" costs, that is, relocation of populace*
- Cost drivers unique to hazardous waste projects per Section 6.3*
- Engineering design costs, if applicable
- Assessment of project contingencies and risk analysis (see Chapter 7)
- Contractor's percentages for overhead and profit (see Chapter 8)

An integral part of any hazardous waste project estimate is the written document termed *basis of estimate*. Particulars covered by the basis of estimate are reviewed at the conclusion of this chapter.

6.1.2 Disclaimer

Data encompassed in tables and examples discussed or detailed within this chapter are hypothetical and are presented for demonstration purposes only. Each estimate prepared for a hazardous waste project is singular, unique, and, as such, most be evaluated on a project-by-project basis.

6.1.3 Recommended Reading

It is strongly recommended that, prior to performing any analysis of known site delays for those items listed in the check list for Section 6.2.3, estimating, cost engineering, environmental, and other industry professionals assigned to preparing, or reviewing an estimate(s) for a hazaroud waste project, review in detail and have a working knowledge and understanding of the latest edition of the *Safety and Health Guidance Manual for Hazardous Waste Site Activities*. This document will be referred to as the Manual within the text of this chapter.

6.2 LABOR PRODUCTIVITY COST DRIVERS

6.2.1 Discussion

To address the need for detailed substantiation in deriving cost drivers, that is, labor productivity multipliers, we will examine significant cost

---

*Discussed within this chapter.

driver variables that need to be identified, defined, and classified when evaluating site labor productivity multipliers used for hazardous waste sites.

6.2.2 Known Site Delays Applicable To A Hazardous Waste Project

The following represents a provisional checklist and synopsis of anticipated labor delays unique to hazardous waste sites:

- Site safety meeting
- Personal protective equipment (PPE) ensemble suitup, resuiting, suit-off for the work day/shift
- Change or replenishment of air tanks
- Work breaks
- Decontamination during the work day/work shift
- Worker cleanup at end of work day/work shift

In this checklist entitled *Known Site Delays Applicable To A Hazardous Waste Project*, all components of assumed site delays affecting the labor work force performing the contracted work scope (i.e., "Direct Labor") must be listed in the delay checklist prepared for *your* project and evaluated. The specific time duration, expressed in minutes, for the item(s) listed in *your* hazardous waste project known site delay checklist forms the foundation for your hazardous waste project labor productivity cost driver multiplier. Time durations used should be ranked on the following criteria:

> *Best*: extracted from previously completed hazardous waste projects for similar levels of worker protection, that is, Levels A, B, C, or D.
> *Good*: based on stopwatch observations for performing the listed operations while in worker protection Levels A, B, C, and D.
> *Poor*: "a best estimate evaluation" for worker protection in Levels A, B, C, and D.

It is recommended that an electronic spreadsheet similar to that shown in Table 6.1 be used when performing this analysis.

6.2.3 Site Safety Meeting(s)

Safety meetings are conducted at the start of the A.M. work shift and supplemented with a refresher, or update, at the conclusion of the noon lunch break. The purpose of these A.M. and P.M. site safety meetings is to discuss and define in detail the hazardous materials to be handled, other safety

Table 6.1 Known Site Delays Applicable to Hazardous Waste Projects

LEVEL OF PROTECTION / TEMPERATURE LEVELS	A <65F (18.3c)	A <85F (29.4c)	A >85F (29.4c)	B <65F (18.3c)	B <85F (29.4c)	B >85F (29.4c)	C <65F (18.3c)	C <85F (29.4c)	C >85F (29.4c)	D <65F (18.3c)	D <85F (29.4c)	D >85F (29.4c)
SITE SAFETY MEETINGS SECTION 6.2.3 — SITE SAFETY MEETINGS	20	20	20	20	20	20	10	10	10	10	10	10
PERSONAL PROTECTIVE EQUIPMENT SECTION 6.2.4 — SUIT-UP/OFF/RE-SUIT(S) FOR THE DAY	60	60	60	60	60	60	40	40	40	10	10	10
WORK BREAKS (TEMP RELATED) SECTION 6.2.6 — MODERATELY COOL [<65F]	60						30			15		
MODERATELY HOT [<85F]		100		40	90			60			30	
HOT [>85F]			140			130			100			50
REPLENISH/REPLACE AIR TANKS SECTION 6.2.5 — MODERATELY COOL [<65F]	80			20			0			0		
MODERATELY HOT [<85F]		120			40			0			0	
HOT [>85F]			130			40			0			0
SECTION 6.2.7 — CLEAN UP AT END OF DAY/SHIFT	20	20	20	20	20	20	20	20	20	20	20	20
SECTION 6.2.8 — TOTAL SITE DELAYS	240	320	370	160	230	270	100	130	170	55	70	90
SECTION 6.2.8 — TOTAL WORK MINUTES	480	480	480	480	480	480	480	480	480	480	480	480
AVAILABLE WORK TIME — TOTAL WORK TIME AVAILABLE	240	160	110	320	250	210	380	350	310	425	410	390
AVAILABLE WORK TIME FACTOR SECTION 6.2.8 — MODERATELY COOL [<65F] / MODERATELY HOT [<85F] / HOT [>85F]	0.50	0.33	0.23	0.67	0.52	0.44	0.79	0.73	0.65	0.89	0.85	0.81
LABOR PRODUCTIVITY FACTOR SECTION 6.2.9 — LIGHT WORK EFFORTS	0.60	0.60	0.60	0.80	0.80	0.80	0.90	0.90	0.90	1.00	1.00	1.00
MODERATE WORK EFFORTS	0.50	0.50	0.50	0.70	0.70	0.70	0.82	0.82	0.82	0.97	0.95	0.95
HEAVY WORK EFFORTS	0.40	0.40	0.40	0.60	0.60	0.60	0.75	0.75	0.75	0.95	0.90	0.90

items, and the level of PPE ensembles (i.e., Level A, B, C, or D) to be worn by worker(s) and support personnel. This PPE level, in turn, dictates the duration of this safety meeting—longest for Level A, shortest for Level D.

6.2.4 Personal Protective Equipment (PPE)

Time constraints for the initial PPE suitup, resuiting, and suitoff during the work day, or work shift, are dependent upon the level of hazardous material exposure. PPE Level A (fully encapsulating, chemical-resistant suit) requires the longest time frame for suiting-up—Level D (minimal protection) requires the shortest. It is recommended that assigned project personnel review in detail Section 8, pp. 8-9 through 8-12 and 8-14 through 8-15 of the Manual.

6.2.5 Changing or Replenishing of Air Tanks

The frequency of replenishing or changing air tanks during a work shift is in direct correlation to the ambient temperatures at the hazardous waste location, coupled with the worker's expenditure of energy (expressed in kilocalories/hr) while performing the assigned work effort (defined as light, moderate, or heavy), in a specific PPE ensemble. Where feasible, air tank(s) replenishment, or changeover, should be accomplished during scheduled site work breaks. It is recommended that assigned project personnel review pp. 8-16 of the Manual.

Depending upon the duration of the hazardous work being performed, and the ambient temperatures at the site—over 85°F (29.4°C) or under 40°F (4.4°C)— particularly in Level A PPE ensemble, alternate methods of providing air to workers in Level A PPE might be a consideration for evaluation.

NOTE: Keep in mind that expenditure of kilocalories/hr while performing work efforts in the assigned worker protection Levels A, B, C, and D is the determining factor for replenishing air tanks while performing light, moderate, or heavy work.

6.2.6 Work Break(s)

Frequent on-site work breaks and monitoring are essential to prevent worker heat stress, reduce worker fatigue, and prevent dehydration while working in the various PPE levels.

NOTE: In geographical locations where work is performed under 40°F (4.4°C), scheduled work breaks are required to prevent worker frostbite and to evaluate worker hypothermia.

The frequency of these on-site work breaks is directly related to the ambient temperatures experienced at the work site while working in the various ensembles—normal or impermeable—and the level of energy (in kilocalories) expended by the worker. Monitoring of the worker by the support staff is accomplished during these on-site work breaks, and includes checking workers for heat stress levels, or identifying possible hypothermal circumstances (abnormal reduction of body heat), ensuring that proper fluids (cold or hot) are provided to prevent either worker heat stress or hypothermal problems, and performing worker decontamination. (See pp. 8-20 through 8-22, Table 8-10 Suggested Frequency of Physiological Monitoring for Fit and Acclimatized Workers, of the Manual.) Depending upon the level of hazardous waste being handled, decontamination of workers can be accomplished during these work breaks.

6.2.7 Cleanup at End of Work Day/Work Shift

Cleanup at the completion of the work day/work shift is mandatory and encompasses the decontamination of the outer PPE garments, and removal, *in specific sequences*, of the protective apparel, and assemblies (gloves, boots, coverall, etc.), with strict adherence to placement of the aforementioned into approved containers for ultimate disposal.

After doffing the outer PPE garments and assemblies, the worker then removes inner clothing coveralls and underwear. These are also deposited into designated containers. The worker than takes the required end-of-work shift shower. Upon completion of this effort, the worker dons street clothes. (See pp. D-2 through D-4 and D-7 through D-13 of the Manual.)

6.2.8 Available Work Time Factor

Summing the project—specific known time delays, expressed in minutes—completes the evaluation for *your* specific hazardous waste site. This summed total is to be deducted from an assumed "standard" 8-hour, or 480-minute, work day/work shift. The remainder, divided by 480 minutes, results in a numerical ratio, designated the *available work time factor*. Table 6.1 depicts a completed example and evaluation of known site delays applicable to hazardous waste projects.

CAUTION: Time durations as indicated are not representative of *your* unique hazardous waste project site.

6.2.9 Labor Productivity Factors

This section reflects *your unique appraisal* of actual, assumed, or estimated labor performance for light, moderate, and heavy work efforts in the various worker protection levels (A, B, C, and D) for the temperature ranges indicated. This is a subjective evaluation and is expressed as a decimal of 100%, for example, a labor productivity factor of 80% would be expressed as 0.80. Again, light, moderate, and heavy work efforts are defined by the expenditure of kilocalories/hr for each assumed ambient temperature range expected. Refer to Table 6.1 which depicts a theoretical assessment of Sections 6.2.2 through 6.2.9 and completes Section 6.2—labor delays that will be encountered on all hazardous waste projects and a technique for evaluating them.

CAUTION: Time durations and labor productivity factors as indicated are not representative of *your* unique hazardous waste project site.

Utilization of an electronic spreadsheet for the evaluation that follows is strongly recommended; it will permit "what if?" studies to be evaluated expeditiously during the preparation of the hazardous waste cost estimate.

6.3 CALCULATION OF LABOR PRODUCTIVITY COST DRIVER MULTIPLIERS

The next element is the resolution of the cost driver labor productivity multipliers, or adjustment factors, for hazardous waste projects, explained in Section 6.3.5 *Net Productivity Ratio*; and the effect on, and interrelationships of, crew size ratio, available work time factor, and assumed labor productivity. Table 6.2 will assist in the understanding of Section 6.3 discussions.

6.3.1 Crew Size and Make Up

Hazardous work crews are made up of "the worker and buddy," plus a support team. The "buddy system" is in effect at all hazardous waste sites (see pp. 9-4 of the Manual). To this is added the unique requirement for hazardous waste sites, that is, a support team. The duties of the support team are safety-related monitoring functions of the "worker and buddy," assisting in changing of air tanks, along with decontamination of personnel and equipment during site work breaks, etc.

Addition of another support team member is a reasonable assumption at hazardous waste sites that will experience ambient temperatures in excess of 85°F (29.4°C) or below 40°F (4.4°C) because of the increased worker

Table 6.2 Calculation of Labor Productivity Multipliers

LEVEL OF PROTECTION / TEMPERATURE LEVELS		A <65F (18.3c)	A <85F (29.4c)	A >85F (29.4c)	B <65F (18.3c)	B <85F (29.4c)	B >85F (29.4c)	C <65F (18.3c)	C <85F (29.4c)	C >85F (29.4c)	D <65F (18.3c)	D <85F (29.4c)	D >85F (29.4c)
CREW SIZE & MAKE-UP SECTION 6.3.1	CREW SIZE (WORKER+BUDDY)	10	10	10	10	10	10	10	10	10	10	10	10
	SUPPORT TEAM	4	4	5	3	3	4	2	2	3	1	1	2
	TOTAL IN CREW	14	14	15	13	13	14	12	12	13	11	11	12
SECTION 6.3.2	CREW SIZE RATIO	0.71	0.71	0.67	0.77	0.77	0.71	0.83	0.83	0.77	0.91	0.91	0.83
AVAILABLE WORK TIME FACTOR SECTION 6.3.3	MODERATELY COOL [<65F]	0.50			0.67			0.79			0.89		
	MODERATELY HOT [<85F]		0.33			0.52			0.73			0.85	
	HOT [>85F]			0.23			0.44			0.65			0.81
LABOR PRODUCTIVITY FACTOR SECTION 6.3.4	LIGHT WORK EFFORTS	0.60	0.60	0.60	0.80	0.80	0.80	0.90	0.90	0.90	1.00	1.00	1.00
	MODERATE WORK EFFORTS	0.50	0.50	0.50	0.70	0.70	0.70	0.82	0.82	0.82	0.97	0.95	0.95
	HEAVY WORK EFFORTS	0.40	0.40	0.40	0.60	0.60	0.60	0.75	0.75	0.75	0.95	0.90	0.90
NET PRODUCTIVITY RATIO SECTION 6.3.5	LIGHT WORK EFFORTS	0.21	0.14	0.09	0.41	0.32	0.25	0.59	0.55	0.45	0.80	0.78	0.68
	MODERATE WORK EFFORTS	0.18	0.12	0.08	0.36	0.28	0.22	0.54	0.50	0.41	0.78	0.74	0.64
	HEAVY WORK EFFORTS	0.14	0.10	0.06	0.31	0.24	0.19	0.49	0.46	0.37	0.76	0.70	0.61
NET PRODUCTIVITY MULTIPLIER SECTION 6.3.6 100 % LABOR COMPONENT	LIGHT WORK EFFORTS	4.67	7.14	11.11	2.44	3.13	4.00	1.69	1.82	2.22	1.24	1.29	1.48
	MODERATE WORK EFFORT	5.56	8.33	12.50	2.78	3.57	4.55	1.85	2.00	2.44	1.28	1.36	1.55
	HEAVY WORK EFFORTS	7.14	10.00	16.67	3.23	4.17	5.26	2.04	2.17	2.70	1.31	1.43	1.64
LABOR AND LABOR & MATERIALS IMPACT MULTIPLIERS SECTION 6.3.7	LIGHT WORK 60% LABOR & 40% MATLS	3.20	4.69	7.07	1.86	2.28	2.80	1.42	1.49	1.73	1.15	1.17	1.29
	LIGHT WORK 50% LABOR & 50% MATLS	2.83	4.07	6.06	1.72	2.06	2.50	1.35	1.41	1.61	1.12	1.14	1.24
	LIGHT WORK 40% LABOR & 60% MATLS	2.47	3.46	5.04	1.58	1.85	2.20	1.28	1.33	1.49	1.10	1.12	1.19
	MODERATE WORK 60% LABOR & 40% MATLS	3.73	5.40	7.90	2.07	2.54	3.13	1.51	1.60	1.86	1.17	1.21	1.33
	MODERATE WORK 50% LABOR & 50% MATLS	3.83	5.50	8.00	1.89	2.29	2.77	1.43	1.50	1.72	1.14	1.18	1.28
	MODERATE WORK 40% LABOR & 60% MATLS	2.82	3.93	5.60	1.71	2.03	2.42	1.34	1.40	1.58	1.11	1.14	1.22
	HEAVY WORK 60% LABOR & 40% MATLS	4.69	6.40	10.40	2.34	2.90	3.56	1.62	1.70	2.02	1.18	1.26	1.38
	HEAVY WORK 50% LABOR & 50% MATLS	4.07	5.50	8.83	2.11	2.58	3.13	1.52	1.59	1.85	1.15	1.22	1.32
	HEAVY WORK 40% LABOR & 60% MATLS	3.46	4.60	7.27	1.89	2.27	2.71	1.42	1.47	1.68	1.12	1.17	1.26
PERSONAL PROTECTIVE EQUIPT COST [PPEC] SECTION 6.3.8	INITIAL $ PURCHASE OF SAME	60	60	80	52	52	60	30	30	40	10	10	20
	DISPOSAL OF GARMENT(S)	20	20	30	18	18	25	18	18	25	6	6	10
	MONITORING EQUIPMENT $'s	15	15	15	10	10	10	7	7	7	7	7	7
	$'s PER DAY/WORKER	95	95	125	80	80	95	55	55	72	23	23	37
	$ COST PER WORK CREW	1330	1330	1875	1040	1040	1330	660	660	936	253	253	444
	$ COST/SHIFT HOUR OR WORK HOUR	11.88	11.88	15.63	10.00	10.00	11.88	6.88	6.88	9.00	2.88	2.88	4.63

site breaks, monitoring, etc. In the data being presented in Table 6.2, a total "productive" crew of 10, that is, five workers and five buddies, was assumed.

6.3.2 Crew Size Ratio

Crew size ratio is the numerical result obtained by dividing the "productive" crew by the total in crew. Therefore, crew size ratio represents the percentage of actual "work hours" in relationship to the makeup of the total crew.

> **Example:** For Level A, temperature(s) 65°F (18.3°C) and light work effort, the crew size ratio would be 5 workers + 5 "buddies" divided by the total crew of 14 (10 "productive" + 4 support team) yielding a crew size ratio of 0.71.

6.3.3 AVAILABLE WORK TIME FACTOR

Summed known site delays, expressed in minutes, divided by the work day (480 minutes), result in the numerical ratio that is designated the *available work time factor*. This is discussed in Section 6.2.8 in Table 6.1, and reproduced again in Table 6.2.

> **Example:** For Level A, temperature(s) 65°F (18.3°C) and light work effort, the available work time factor would be total work hours, expressed in minutes, less total summed site delay minutes, divided by total work day minutes, that is, $(480 - 240)/480 = 0.50$.

6.3.4 Labor Productivity Factors

This is developed in Section 6.2.9, Table 6.1, and reproduced again in Table 6.2 with labor productivity factors expressed as a decimal of 100%.

6.3.5 Net Productivity Ratio

The cumulative multiplication of the labor productivity factor from Section 6.2.9 (expressed as a decimal of 100%) times the crew size ratio (see Section 6.3.2) times the available work time factor developed in Section 6.2.8 yields the net productivity ratio.

> **Example:** For Level A, temperature(s) 65°F (18.3°C) and light work effort, this cumulative evaluation would be $(0.60) \times (0.71) \times (0.50) = 0.21$.

This calculation is then repeated for *each* personal protective equipment level—light, moderate, and heavy work effort—and for *each* temperature range expected; hence the recommendation for the utilization of an electronic medium to perform the required calculations.

> **Example**: For Level A, temperature(s) 65°F (18.3°C) and light work effort, the net productivity ratio would be (0.50) × (0.71) × (0.60) = 0.21.

> **Example**: For Level A, temperature(s) 65°F (18.3°C) and moderate work effort, the net productivity ratio would be (0.50) × (0.71) × (0.50) = 0.18.

> **Example:** For Level A, temperature(s) 65°F (18.3°C) and heavy work effort, the net productivity ratio would be (0.50) × (0.71) × (0.40) = 0.14.

6.3.6 Net Productivity Multiplier

Net productivity multiplier is the reciprocal of the net data calculated above in Section 6.3.5 for light, moderate, and heavy work efforts, and again, for each PPE level and temperature range expected:

> **Example**: For Level A, temperature(s) 65°F (18.3°C) and light work effort, the net productivity multiplier is the reciprocal (1) divided by 0.21 = 4.67.

> **Example**: For Level A, temperature(s) 65°F (18.3°C) and moderate work effort, the net productivity multiplier is the reciprocal (1) divided by 0.18 = 5.56.

> **Example**: For Level A, temperature(s) 65°F (18.3°C) and heavy work effort, the net productivity multiplier is the reciprocal (1) divided by 0.14 = 7.14.

6.3.7 Labor and Labor/Materials Impact Multipliers

Two cost driver components must be evaluated. The first component is for 100% labor operations, (for example, equipment operator(s)); the other component to be evaluated is a combination of labor and materials percentages. Therefore, respective multipliers for all labor and a combination of labor and materials percentages must be considered, as outlined in the following examples:

Example: For Level A, temperature(s) 65°F (18.3°C), light work effort, and a 100% labor-intensive operation the labor impact multiplier would be (4.67 × 1.00) = 4.67.

Example: For Level A, temperature(s) 65°F (18.3°C), moderate work effort, and a 100% labor-intensive operation the labor impact multiplier would be (5.56 × 1.00) = 5.56.

Example: For Level A, temperature(s) 65°F (18.3°C) heavy work effort and 100% labor-intensive operation the labor impact multiplier would be (7.14 × 1.00) = 7.14.

However, utilizing the same temperature level of 65°F (18.3°C) and work effort level as outlined above, but with a labor component of 60% and material component of 40%, this multiplier would be calculated as follows:

Light labor effort:
Labor component + material component
[(4.67 × .60) + .40] = 3.20
Moderate labor effort:
Labor component + material component
[(5.56 × .60) + .40] = 3.73
Heavy labor effort:
Labor component + material component
[(7.14 × .60) + .40] = 4.69

Similar labor and material multipliers for various combinations of labor and material ratios commencing with 60% labor and 40% materials for light, moderate, and heavy work efforts, respectively, can be calculated (see Table 6.2).

Following the above examples, Table 6.3 can be prepared depicting labor productivity multipliers derived for light, moderate, and heavy worker effort classifications for combinations of labor and material ratios: 100% labor (0% materials), decreasing to 20% labor (80% materials).

6.3.8 Personal Protection Equipment Costs (PPEC) Per Shift Hour

Cost drivers for personal protection equipment costs per shift hour are:

- Level of personal protection required (A, B, C, or D)
- Frequency of decontamination(s) per work day/shift

Table 6.3 Calculation of Various Labor and Material Component Multipliers

LEVEL OF PROTECTION / TEMPERATURE LEVELS	A <65F (18.3c)	A <85F (29.4c)	A >85F (29.4c)	B <65F (18.3c)	B <85F (29.4c)	B >85F (29.4c)	C <65F (18.3c)	C <85F (29.4c)	C >85F (29.4c)	D <65F (18.3c)	D <85F (29.4c)	D >85F (29.4c)
CREW SIZE & MAKE-UP												
CREW SIZE (WORKER+BUDDY)	10	10	10	10	10	10	10	10	10	10	10	10
SUPPORT TEAM	4	4	5	3	3	4	2	2	3	1	1	2
TOTAL IN CREW	14	14	15	13	13	14	12	12	13	11	11	12
CREW SIZE RATIO	0.71	0.71	0.67	0.77	0.77	0.71	0.83	0.83	0.77	0.91	0.91	0.83
AVAILABLE WORK TIME FACTOR												
MODERATELY COOL [<65F]	0.50			0.67			0.79			0.89		
MODERATELY HOT [<85F]		0.33			0.52			0.73			0.85	
HOT [>85F]			0.23			0.44			0.65			0.81
LABOR PRODUCTIVITY FACTOR												
LIGHT WORK EFFORTS	0.60	0.60	0.60	0.80	0.80	0.80	0.90	0.90	0.90	1.00	1.00	1.00
MODERATE WORK EFFORTS	0.50	0.50	0.50	0.70	0.70	0.70	0.82	0.82	0.82	0.97	0.95	0.95
HEAVY WORK EFFORTS	0.40	0.40	0.40	0.60	0.60	0.60	0.75	0.75	0.75	0.95	0.90	0.90
NET PRODUCTIVITY RATIO												
LIGHT WORK EFFORTS	0.21	0.14	0.09	0.41	0.32	0.25	0.59	0.55	0.45	0.80	0.78	0.68
MODERATE WORK EFFORTS	0.18	0.12	0.08	0.36	0.28	0.22	0.54	0.50	0.41	0.78	0.74	0.64
HEAVY WORK EFFORTS	0.14	0.10	0.06	0.31	0.24	0.19	0.49	0.46	0.37	0.76	0.70	0.61
NET PRODUCTIVITY MULT (100 % LABOR COMPONENT)												
LIGHT WORK EFFORTS	4.67	7.14	11.11	2.44	3.13	4.00	1.69	1.82	2.22	1.24	1.29	1.48
MODERATE WORK EFFORT	5.56	8.33	12.50	2.78	3.57	4.55	1.85	2.00	2.44	1.28	1.36	1.55
HEAVY WORK EFFORTS	7.14	10.00	16.67	3.23	4.17	5.26	2.04	2.17	2.70	1.31	1.43	1.64
% LABOR & MATLS COMBINATIONS — LIGHT WORK EFFORTS												
LIGHT WORK 100% LABOR	4.67	7.14	11.11	2.44	3.13	4.00	1.69	1.82	2.22	1.24	1.29	1.48
LIGHT WORK 90% LABOR & 10% MATLS	4.30	6.53	10.10	2.30	2.92	3.70	1.62	1.74	2.10	1.22	1.26	1.43
LIGHT WORK 80% LABOR & 20% MATLS	3.94	5.91	9.09	2.15	2.70	3.40	1.55	1.66	1.98	1.19	1.23	1.38
LIGHT WORK 70% LABOR & 30% MATLS	3.57	5.30	8.08	2.01	2.49	3.10	1.48	1.57	1.85	1.17	1.20	1.34
LIGHT WORK 60% LABOR & 40% MATLS	3.20	4.68	7.07	1.86	2.28	2.80	1.41	1.49	1.73	1.14	1.17	1.29
LIGHT WORK 50% LABOR & 50% MATLS	2.84	4.07	6.06	1.72	2.07	2.50	1.35	1.41	1.61	1.12	1.15	1.24
LIGHT WORK 40% LABOR & 60% MATLS	2.47	3.46	5.04	1.58	1.85	2.20	1.28	1.33	1.49	1.10	1.12	1.19
LIGHT WORK 30% LABOR & 70% MATLS	2.10	2.84	4.03	1.43	1.64	1.90	1.21	1.25	1.37	1.07	1.09	1.14
LIGHT WORK 20% LABOR & 80% MATLS	1.73	2.23	3.02	1.29	1.43	1.60	1.14	1.16	1.24	1.05	1.06	1.10
MODERATE WORK EFFORTS												
MODERATE WORK 100% LABOR	5.56	8.33	12.50	2.78	3.57	4.55	1.85	2.00	2.44	1.28	1.36	1.55
MODERATE WORK 90% LABOR & 10% MATLS	5.10	7.60	11.35	2.60	3.31	4.20	1.77	1.90	2.30	1.25	1.32	1.50
MODERATE WORK 80% LABOR & 20% MATLS	4.65	6.86	10.20	2.42	3.06	3.84	1.68	1.80	2.15	1.22	1.29	1.44
MODERATE WORK 70% LABOR & 30% MATLS	4.19	6.13	9.05	2.25	2.80	3.49	1.60	1.70	2.01	1.20	1.25	1.39
MODERATE WORK 60% LABOR & 40% MATLS	3.74	5.40	7.90	2.07	2.54	3.13	1.51	1.60	1.86	1.17	1.22	1.33
MODERATE WORK 50% LABOR & 50% MATLS	3.28	4.67	6.75	1.89	2.29	2.78	1.43	1.50	1.72	1.14	1.18	1.28
MODERATE WORK 40% LABOR & 60% MATLS	2.82	3.93	5.60	1.71	2.03	2.42	1.34	1.40	1.58	1.11	1.14	1.22
MODERATE WORK 30% LABOR & 70% MATLS	2.37	3.20	4.45	1.53	1.77	2.07	1.26	1.30	1.43	1.08	1.11	1.17
MODERATE WORK 20% LABOR & 80% MATLS	1.91	2.47	3.30	1.36	1.51	1.71	1.17	1.20	1.29	1.06	1.07	1.11
HEAVY WORK EFFORTS												
HEAVY WORK 100% LABOR	7.14	10.00	16.67	3.23	4.17	5.26	2.04	2.17	2.70	1.31	1.43	1.64
HEAVY WORK 90% LABOR & 10% MATLS	6.53	9.10	15.10	3.01	3.85	4.83	1.94	2.05	2.53	1.28	1.39	1.58
HEAVY WORK 80% LABOR & 20% MATLS	5.91	8.20	13.54	2.78	3.54	4.41	1.83	1.94	2.36	1.25	1.34	1.51
HEAVY WORK 70% LABOR & 30% MATLS	5.30	7.30	11.97	2.56	3.22	3.98	1.73	1.82	2.19	1.22	1.30	1.45
HEAVY WORK 60% LABOR & 40% MATLS	4.68	6.40	10.40	2.34	2.90	3.56	1.62	1.70	2.02	1.19	1.26	1.38
HEAVY WORK 50% LABOR & 50% MATLS	4.07	5.50	8.84	2.12	2.59	3.13	1.52	1.59	1.85	1.16	1.22	1.32
HEAVY WORK 40% LABOR & 60% MATLS	3.46	4.60	7.27	1.89	2.27	2.70	1.42	1.47	1.68	1.12	1.17	1.26
HEAVY WORK 30% LABOR & 70% MATLS	2.84	3.70	5.70	1.67	1.95	2.28	1.31	1.35	1.51	1.09	1.13	1.19
HEAVY WORK 20% LABOR & 80% MATLS	2.23	2.80	4.13	1.45	1.63	1.85	1.21	1.23	1.34	1.06	1.09	1.13

- Number of resuitings required in a work day/shift
- Hours worked per work day/shift

Table 6.2 shows an assumed cost for the following:

1. Initial purchase of the PPE: $60
2. Disposal of the PPE ensemble: $20
3. Pro rating of the monitoring equipment: $15
4. Assumed crew shift of 8 hours: 112

Example: Per Table 6.2, costs per crew work day for Level A PPE and temperature of 65°F (18.3°C) would be derived as follows:

Total number in work crew multipled by PPE cost per worker divided by total crew work hours produces a dollar cost per shfit hour for PPE ensembles.

Example: Total crew size 14 × $95/PPEC per worker ÷ by total crew work hours (14 in crew × $95/worker) ÷ by 112 crew work hours) = $11.88/shift work hour.

6.4 OTHER COST DRIVERS UNIQUE TO HAZARDOUS WASTE PROJECTS

6.4.1 Other Cost Drivers Unique to Hazardous Waste Projects

The following encompasses additional unique cost drivers for hazardous waste projects. Should any of these singular cost drivers be omitted, ignored, not evaluated, or assessed incorrectly, a significant cost risk will have been introduced into the completed cost estimate, creating an environment whereby a contract undertaken at a hazardous waste project would require a large allowance/contingency so as not to exceed budgetary maximums. Significant efforts to meticulously evaluate the cost drivers for the following items would be a prudent investment by a company.

- Medical examinations
- Worker decontamination stations
- Shower facility and change house(s)
- Vehicle/equipment/tools decontamination station(s)
- Disposal of contaminated wastewater

- Personal Protection Equipment Costs (PPEC)
- Testing and monitoring equipment
- Site communication system(s)
- Increased first aid facilities
- Administration of hazardous waste site projects
- Field laboratory setup and operation
- Site security, and perimeter fencing
- Field design engineering support during remedial efforts
- Hazardous waste project permits
- Environmental due diligence
- On-site dust mitigation
- Community and public relations

6.4.2 Medical Examinations

Each worker, field staff member, field supervisor, and all office personnel assigned to a hazardous waste project, must have a pre-employment
medical examination prior to working on, or entering, the hazardous project work area or site. The rationale for these examinations is to establish
a medical baseline for all project personnel prior to their commencement
of work. When there is an employee or worker reduction in force, a termination of employee, or an employee/worker who quits prior to completion of the project, there is an additional requirement for an exit medical
examination. Cost considerations must be allowed for those in the work
force (worker and staff) who, for a myriad of reasons, are examined and
do not pass the screening requirements set for employment on hazardous
waste projects; normal labor attrition is also expected. Medical costs associated with all candidates screened, whether they pass or fail, will accrue
to your project.

The cost for these medical examinations varies. Depending upon the
locale, preliminary work screening and postwork medical examinations
could run the cost gamut from $200 to $600 or *more* per medical examination *per person*. Keep in mind that not all persons being examined will
pass the screening criteria established. This singular area must be judi-
ciously evaluated.

6.4.3 Worker Decontamination Station(s)

Requirements for these facilities are site specific and contingent upon the
project's time duration (days, weeks, months, year(s)), coupled with the

work scope to be accomplished at the hazardous waste site and the level of worker protection required. The facility provided could be temporary, or permanent; for example, a simple outside deluge shower and hence minimal costs, or a permanent enclosed multihead deluge system with underground collection systems and hence significant costs. All contaminated wastewater from these worker decontamination stations must be collected, removed from the site, and disposed of.

6.4.4 Shower Facility and Change House

Workers are required to shower upon completion of a work shift before changing into street clothes. Again, the duration of the work efforts at the hazardous waste site determines the economics to be employed in setting up either a shower trailer or a combined shower area and change house. Effluent from these facilities also has to be collected, removed from the site, and disposed of.

6.4.5 Vehicle/Equipment/Tools Decontamination Station(s)

Every vehicle, all equipment, and all work tools entering the area of a hazardous waste site must be decontaminated prior to leaving the physical boundaries of the hazardous waste site; to wit: if the work effort relates to capping of an existing site, then every truck bringing in the capping materials is affected. This requirement, that is, decontamination of vehicles and equipment prior to leaving a hazardous waste project, encompasses not only construction vehicles used within the contaminated area; it could, under specific conditions, also involve the personal vehicles of the workers, delivery trucks, and vehicles of the site and security staff as well.

Effluent from the vehicle decontamination station(s) must be collected, removed from the site, and disposed of. The labor component for operation of a vehicle and equipment decontamination station(s) is *in addition* to the decontamination efforts discussed in Section 6.2.8. Labor to accomplish this end of shift cleanup must be included in the overall hazardous waste project cost estimate. Again, the effluent from decontamination efforts for equipment and tools must be collected, removed offsite, and disposed of.

6.4.6 Disposal of Contaminated Wastewater

The collection and ultimate disposal of contaminated waste water from the various vehicle/equipment/tool decontamination station(s), shower and change house facilities, and field decontamination facilities is a cost-sensitive consideration. This is not a cost evaluation to be taken lightly, because there are special collection and storage system requirements. Costs associated with these collection systems, and the daily, weekly, or monthly on-site storage facilities prior to collection, transportation, treatment, and disposal of this contaminated wastewater, (classed as a hazardous waste), can be significant.

Do not fail to fully evaluate the additional staff cost elements associated with permits for transportation of hazardous waste water materials and disposal of same at designated sites; these also add to the hazardous waste project estimate.

6.4.7 Personal Protection Equipment Costs (PPEC)

The primary cost driver(s) for the total costs for this facet of a cost estimate is a function of the total work hours developed for the contracted work scope for the hazardous waste project based on the details and analysis of Section 6.3.8. Other significant cost drivers are as listed in Section 6.3.8, (i.e., initial costs of the components of worker protection Levels A, B, C, and D and the classification of the hazardous waste project), plus monitoring instrumentation and additional resuitings of the worker during the work day or work shift. Disposal of contaminated personnel equipment items placed into designated hazardous waste containers per Section 6.2.8 represents additional cost allowances/considerations.

6.4.8 Testing and Monitoring Equipment

Testing and monitoring requirements are site specific, and defined by the Request for Quote (RFQ) or Request for Proposal (RFP). Cost drivers for this line item are based on testing and monitoring equipment costs, plus supporting personnel to install, operate, and maintain the testing and monitoring equipment sites or locations. Based on certain geographical areas, there would be a specific requirement for an equipment shelter. Do not overlook power requirements and associated electrical connections and utility costs incurred during the life of your project.

6.4.9 Site Communication System(s)

Again, a site-specific evaluation: the cost driver for this item is *time*. This site communication system is in *addition* to the "customary" on-site walkie-talkie sets commonly used on a construction site. Keep in mind that you're involved with remedial activities that are associated with an existing hazardous waste site that is in someone's "backyard." Public, local, state, and governmental agencies must be advised of any "incidents" that occur on your hazardous waste site *before* your neighbor calls *his* local, state, or governmental representative—surprises or environmental ambushes should be avoided.

Evaluate the number of additional mobile installations as well as "fixed installations" for local, state, private, and governmental agencies. Do not underestimate rental costs or omit power requirements for same, related equipment elements (antennas, grounding, etc.) and associated utility costs expected to be incurred during the life, (i.e., time frame), of your project.

6.4.10 Increased First Aid Facilities

Always bear in mind you are working on a hazardous waste project; a worker exposed to a hazardous material "incident" has to be treated/remedied *now*! The "normal" first aid construction facility manned by Red Cross certified personnel may not be applicable for your hazardous waste project. There is a high probability that the first aid facility required at your project will have to be manned by professional medical personnel (i.e., a licensed nurse and doctor). Recommend that the evaluation of this cost driver be assigned to an estimating specialist/outside consultant firm that has experience handling this type of evaluation, because the medical equipment, supplies, and supporting facility must be analyzed fully to present a fair estimate of this added requirement. Do not omit utility costs for this facility, nor the costs for the treatment and disposal of medical waste, as these are hazardous waste materials and must be treated and disposed of accordingly.

6.4.11 Administration of Hazardous Waste Site Project(s)

Fully evaluate space requirements and the type of facility (trailers, module assemblies, etc.) to house all of the staff, management functions, and specialty subcontractors and consultant operations. Added costs for a required facility for the administration of hazardous waste projects should be based on a "normal" construction project of similar cost magnitude. This approach will assist in identifying the additional field office areas and spaces that are essential for your hazardous waste project office.

Efforts associated with permits, added site communications systems, collection and disposal of hazardous wastewater, laboratory and site monitoring data, governmental reports, specialty subcontractors and consultant operations, site security administration, etc., all have one thing in common: *added field office staff.* Added field office staff costs for the administration of hazardous waste projects should be based on a comparison of "normal staffing" for a construction project of similar cost magnitude. This approach will assist in identifying additional field office skills now required to man the hazardous waste project office. Skilled technical personnel may require mobilization and associated relocation costs.

Do not omit the additional costs associated with office equipment (desktop computers, laser printers, telephones, fax machines, etc.) to support the administrative efforts of the increased staff. Increased fax telephone costs should not be overlooked as governmental agencies seem to favor this particular form of communication.

6.4.12 Field Laboratory Setup And Operation

The field laboratory operation could be an ancillary area of the testing and monitoring phase of the contracted work scope. Again, this is site specific as to the type of field laboratory tests to be performed on, and for, the assigned hazardous waste project. Cost drivers are laboratory equipment, and the type of tests to be performed (average or complex), plus skilled laboratory personnel to perform same (for example, "high" or "low" tech, frequency of tests and type (simple or complex), including "off site" costs for tests that cannot be performed on site.

Evaluation of this cost driver should be assigned to an estimating specialist/outside consultant firm that has experience handling this type of evaluation, because the field laboratory equipment, supplies, and supporting facility must be analyzed fully to present a fair estimate of this added requirement. Do not omit utility costs for this facility, nor costs associated with the disposal of field laboratory test specimens, as these are hazardous waste materials and must be treated and disposed of accordingly.

6.4.13 Site Security and Perimeter Fencing

Site Security

Site security cost drivers are the numbers of security personnel required per shift, plus associated vehicles, communications systems, and facilities for same. This is a 24-hour, three-shift work assignment, 7 days a week, including holidays, for the duration of the hazardous waste project. This is a significant cost for a long-term hazardous waste site project.

These personnel are subject to medical examinations at the start and completion of the contracted work assignments. Review other costs relating to decontamination of vehicles used on site for the waste project; again, the effluent from vehicle and tool decontamination must be collected, removed offsite, and disposed of.

Perimeter Fencing

Cost drivers for perimeter fencing encompass degree of site clearing and necessary site prep, plus lineal foot and type of fence required (two or three strand barbed wire at top of fence), per project specifications, for the site security perimeter fencing. Additional drivers could be a site requirement for illumination of the fenced perimeter area, and a requirement for the lower portion of the fence fabric to be embedded in concrete, thus preventing egress under the fence fabric. Salvage of the fencing components (line posts, top rail, fence fabric and barbed wire) at the end of the hazardous waste project is possible, unless the security fence is to remain in place, or the contents of the hazardous waste project prevent the fencing elements from being reused.

6.4.14 Field Design Engineering Support During Remedial Efforts

Primary cost drivers for field design engineering support are numbers of assigned personnel, technical skill levels, monthly salaries, relocation packages, or temporary assignments, and the assumed duration in weeks, months, and years on site. The assumed on-site duration would match that of your hazardous waste project. Other cost drivers would be field office space requirements, computer equipment, increased field office staff to support design engineering efforts, vehicles for on-site usage, training of design engineering personnel for use of PPE required for on-site investigations.

6.4.15 Hazardous Waste Project Permits

Hazardous waste project permits represent an area of considerable cost currently not fully understood or evaluated by those preparing hazardous waste project cost estimates. Historically, costs associated with preparing and obtaining these construction site permits have been submerged within the total field staff or other costs. To fully evaluate *all* associated costs, each permit required for the hazardous waste project should be identified, listed by project/construction management, and placed on a computer timeline schedule; costs drivers should be evaluated for:

- Preparation of the permit documents
- Resubmittal of the permit documents
- Maintaining compliance with permits
- Monitoring to verify/assess compliance
- Future expansions/modifications of the permit base
- Closure of the hazardous waste site upon completion

All of the above are cost sensitive to time durations and numerous resubmittals, for whatever cause, to effect issuance of these permits, and are habitually overlooked. Unfortunately, project/construction management frequently assigns diverse and uncoordinated segments of the staff to follow the permitting process. This can result in a large expenditure of valuable, and hence, costly resources. It is recommended that a "compact and coordinated special permit team" be utilized for your project and costed out accordingly.

Because of the permit bureaucracy, there are numerous "fingers in the pie" (federal, state, local, and city departments) who are involved in the permitting process; costs to accomplish this permitting process must be carefully monitored to ensure value for time (work hours) expended and adherence to budgetary constraints.

6.4.16 Environmental Due Diligence

A relatively new area for evaluation of environmental cost assessment has evolved into a special discipline called *environmental due diligence investigation*. Practitioners of this new discipline are registered or certified to attest to their qualifications to conduct the required investigations. This new area could entail a substantial cost exposure associated with your project in this area and should not be ignored.

6.4.17 On-Site Dust Mitigation

The cost driver for on-site dust mitigation is an interesting area. If too much water is applied to control dust from the hazardous waste project, there will be extensive runoff from the watered-down surfaces. Where does this runoff drainage go? Remember, any runoff or drainage from this site is classed as hazardous waste material. Was this evaluated in the volume of water to be collected? (See section 6.4.6.) The other side of the coin is when not enough water is used and dust clouds rise from the site. This leads us to another overlooked area, which, when approached diplomatically and properly, can be one of the most effective mitigation endeavors for a hazardous waste project.

6.4.18 Community and Public Relations

Cost drivers for this component are site specific as to the remedial action for the hazardous waste materials currently in situ within the boundaries of the project. Inhabitants of areas in close proximity, particularly single or multifamily dwellings, schools, hospitals, etc., all come together and discuss one area of common interest. *What's going on "over there?"* "Over there" consists of uninformed discussions of your project during weekend cookouts, evening dinners, PTA meetings, and at the local supermarket. Again, keep in mind that you're involved with remedial activities associated with an existing hazardous waste site located in someone's "back yard;" your project management knows whats going on, so why shouldn't others in the surrounding area also be privy to the same information?

As with other specialty areas discussed, it is recommended that this public relations cost evaluation be assigned to a consultant who specializes in public relations. Significant rewards to your company will accrue for the project from a well-prepared, well-presented public relations program, and is a prudent item for a company to investigate.

6.5   ELECTRONIC MEDIA

6.5.1 Advantages of Electronic Spreadsheets

The singular advantage of using electronic spreadsheets in the preparation of complex hazardous waste project cost estimates lies in the electronic media's ability to rapidly calculate and sum cost components. Additionally, this electronic medium results in a minimum turnaround time to evaluate "what-if" scenarios. Adoption of electronic spreadsheets allows evaluation of the singular significant cost driver, which, for hazardous waste projects, is the assumed labor productivity of the work force

in various worker protection levels. Those required calculations to evaluate "what-if" studies relative to Tables 6.1, 6.2 and 6.3 of this chapter,
and minute variations in Section 6.2.10 (Labor Productivity Factors) and
Section 6.3.5 (Net Productivity Ratios) are extensive, even with the assistance of these electronic spreadsheets.

6.5.2 Tentative Spreadsheet Formats

Other spreadsheet formats are shown at the end of Chapter 5. Examples
of suggested spreadsheet formats for a leachate collection system and
building asbestos removal are shown in Tables 6.4 and 6.5.

Table 6.4   Leachate Collection System: Example Spreadsheet

	Qty. Unit	Labor	Materials	Sub-Cont.	Totals
Clear and grub	Acre			XXXX	XXX
Drill collection well(s)	Each			XXXXX	XXXXXX
Mobilization/demobilization					
Spot drill rig	Estimate details				
Raise mast and Drill xx lft					
Lower mast and move					
Load/remove drill materials	Month	XXXX	XXXX	XXXX	XXX
Leachate collection system	LF			XXXXX	XXXXXX
Fab PVC pipe system					
Fab PVC pipe well trees					
Install trees and collection systems	Estimate details				
Install pipe supports					
Install drain lines					
Storage tanks					
System test					
Install collecting pumps	Each	XXXX	XXXX	XXXX	XXXXXX
Concrete Fdns for same					
Install/align/grout					
Electrical Connections					
Install electrical system	LS			XXXXX	XXXXXX
Dust mitigatiom	Month	XXXX	XXXX		XXXXXX
Maintain const roads	Month	XXXX	XXXX		XXXXXX

Table 6.4 (*continued*)

	Qty. Unit	Labor	Materials	Sub-Cont.	Totals
Runoff control and ditches	LF	XXXX	XXXX		XXXXXX
Maintain ditches	Month				
Decon const vehicles	Each	XXXX	XXXX		XXXXXX
Decon const tools	Month	XXXX	XXXX		XXXXXX
Perimeter fencing	LF	XXXX	XXXX		XXXXXX
Scope growth	%				XXXXXX
Bid allowance	%				XXXXXX
			Subtotal Construction		$XXXXX
Health and safety training	Each	XXXX	XXX		XXXX
Site security measures	Month	XXXX	XXX		XXXX
Worker decon stations	Month	XXXX	XXX		XXXX
Equipment decon station	Month	XXXX	XXX		XXXX
Staff/trade/delivery decon	Month	XXXX	XXX		XXXX
Prework medical exams	Each	XXXX		XXXX	XXXX
Postwork medical exams	Each	XXXX		XXXX	XXXX
First aid station	Month	XXXX		XXXX	XXXX
Communications systems	Month	XXXX		XXXX	XXXX
Purchase disposal PPE clothing	Each	XXXX	XXXX		XXXX
Disposal of clothing	Each	XXXX	XXXX		XXXX
Disposal effluent water	Month	XXXX	XXXX		XXXX
DOT transport permits	Each	XXXX		XXXX	XXXX
Hazardous waste site permits	Each	XXXX	XXXX		XXXX
Haz. waste disposal fee	Load			XXXX	XXXX
Other Costs as applicable (see Sections 6.4.11-12, 6.4.14, 6.4.16, 6.14.18)		XXXX	XXXX	XXXX	XXXX
	Subtotal Health and Safety				$XXXXXXX
Subtotal Implementation Costs	Construction Plus Health and Safety				$XXXXXXX
Engineering Design Costs	Percentage of construction cost				$XXXX
Grand Total	Leachate Collection System				$XXXXXXX

Table 6.5 Building Asbestos Removal: Example Spreadsheet

	Qty. Unit	Labor	Materials	Sub-Cont.	Totals
Vacuum Work Area	SF			XXXX	XXXX
Drap Work Area	SF			XXXXXX	XXXXXX
Plastic Drap Floor Area					
Plastic Drap Wall Areas	Estimate				
Plastic Drap Ceiling Area	details				
Bag & Remove Asbestos Matls					
Vacuum Completed Work Area	SF			XXXX	XXXX
Instal Ventl Syst. and Maintain	Month	XXXX	XXXX	XXXX	XXXX
Instal Water Syst. and Maintain	LS			XXXX	XXXX
Disposal of Asbestos Matls.	Ton Mile			XXXX	XXXXX
Haz. Waste Disposal Fee	Load			XXXX	XXXX
Asbestos Dust Mitigation	Month	XXXX	XXXX		XXXXX
Decon. Const. Vehicles	Each	XXXX	XXXX		XXXXX
Decon. Const. Tools	Month	XXXX	XXXX		XXXXX
Scope Growth	%				XXXXX
Bid Allowance	%				XXXXX
				Subtotal Construction	$XXXXX
Health and safety training	Each	XXXX	XXX		XXXX
Site security measures	Month	XXXX	XXX		XXXX
Worker decon. stations	Month	XXXX	XXX		XXXX
Equipment decon. station	Month	XXXX	XXX		XXXX
Staff/trade/delivery Decon.	Month	XXXX	XXX		XXXX
Prework medical exams	Each	XXXX		XXXX	XXXX
Postwork medical exams	Each	XXXX		XXXX	XXXX
First aid station	Month	XXXX		XXXX	XXXX
Purchase disposal PPE clothing	Each	XXXX	XXXX		XXXX
Disposal of clothing	Each	XXXX	XXXX		XXXX
Disposal of asbestos water	Month	XXXX	XXXX		XXXX

Table 6.5 (*continued*)

	Qty. Unit	Labor	Materials	Sub-Cont.	Totals
DOT transport permits	Each	XXXX		XXXX	XXXX
Haz. waste site permits	Each	XXXX	XXXX		XXXX
Other Costs as applicable (see Sections 6.4.16, 6.4.18)		XXXX	XXXX	XXXX	XXXX
	Subtotal Health and Safety				$XXXXXXX
Subtotal implementation costs	Construction plus Health and Safety				$XXXXXXX
Engineering design costs	Percentage of construction cost				$XXXX
Grand Total	Building Asbestos Removal				$XXXXXX

The items listed are variously applicable for the city, building configuration, height, local codes, etc.

6.6 EXAMPLE USING DEVELOPED LABOR PRODUCTIVITY FACTORS

6.6.1 Use of Developed Labor Productivity Factors for Scope Estimates

An example of the theoretical application of a labor productivity cost driver developed using *your company data base* is presented at the end of this chapter. (See Appendix 6A).

6.7 ADDITIONAL SOURCES OF INFORMATION

The following is a listing of recommended areas for additional cost information relative to hazardous waste estimates, or specific areas discussed within this chapter.

6.7.1 Your Company's Historical Cost Records

These are an excellent source of cost data or information. These can be analyzed, categorized, and computerized into an electronic cost data base applicable for use with ease for current or future cost estimating work efforts relative to hazardous waste projects.

6.7.2 Your Individual Cost Studies or Special Studies

Update the individual cost/special studies that you have assembled over the years and perform the same analysis as outlined in the preceding section. Include the results of this analysis in your company data base. Better yet, share this information with other cost professionals by publishing a technical paper.

6.7.3 Obtain The Cost Data Needed From The Field

Do you need actual field time data for suitup, resuiting, suitoff in worker Level protection B, or for end of work shift cleanup? Ask your field cost representative to perform a series of cost studies on same. Compare actual time durations with the time durations you utilized in preparing Table 1 for your unique hazardous waste project (see Sections 6.2.4 and 6.2.7).

6.7.4 Governmental Publications

An area that had not been fully documented at the time this chapter was prepared was the productivity performance of combat individuals/teams while in fully encapsulated gasproof assemblies used by the U.S. armed forces (and others) during the "Desert Storm" operations. Daylight ambient desert temperatures experienced by combat personnel performing various tasks/operations while in these fully encapsulated ensembles approximated worker protection Level A required on hazardous waste projects. It is possible that in the near future information relative to fluid(s) intake, rest periods, and monitoring data associated with varying work levels (light, moderate, and heavy) based on (combat) deployments associated with Operation Desert Storm could be the subject of a publication(s) by one or more of the U.S. armed forces' medical branches. The reader should be cognizant of this probability.

However, the following publications do exist: *Occupational Safety and Health Guidance Manual for Hazardous Waste Site Activities* and *Cost Estimating Handbook for Environmental Restoration*.

6.7.5 AACE International Publications or Technical Papers

Some of these are: "Cost Control in the 1990s"—Getting the Project Started Right, Second Annual Skills & Knowledge Seminar Proceedings Sponsored by Rocky Mountain Section AACE, Inc., October 23, 1991. Selected technical papers:

Project Scope Definitions—A Practical Approach
Cost Aspects of Mining Regulations
Development of Environmental Permitting Programs
Reclaiming and Bonding For The Minerals Industry

For further information, contact AACE International, P.O. Box 1557, Morgantown, WV 26507–1557. Phone (304) 296–8444 or toll-free 800–858–COST.

6.8   UTILIZING THIS CHAPTER ON ACTUAL PROJECTS

6.8.1 Applying the Concepts in This Chapter to Actual Projects

It is postulated that the identification of unique cost drivers, coupled with a structured approach to preparing cost assessments and estimates for hazardous waste projects, as outlined in Chapter 6 of this publication, will result in a definitive cost audit trail permitting cost professionals assigned to a hazardous waste project to "sample" various time durations, and units of production assumed or utilized, in the preparation of a completed and approved hazardous waste project cost estimate. An example of this "sampling" approach was tentatively outlined and discussed in Section 6.7.3. Based on "feedback" from the field, project cost professionals compare time/cost components of the "as planned" to "field actuals," identifying labor workhours and dollar cost variants. These work hours and cost variants, positive or negative, signal project labor overruns/ underruns, as well as implying associated schedule impact(s) to the project. This singular comparison, coupled with other "sampling" programs performed on a regular basis (daily or weekly) over the construction period of the hazardous waste project permits cost professionals to accomplish what we in this profession call *cost and schedule visibility*. Cost and schedule visibility, coupled with a cost trending program, provides early flags for project management corrective action of budgetary overruns or schedule extensions, with a derived result that all cost professionals continuously strive for—*no cost or schedule surprises for their project*.

Basis of Estimate

As a final assist in attaining the desired goal of no cost or schedule surprises for a hazardous waste project, a written document called the *Basis of Estimate* is prepared. This document, with its detailed crew cost per work hour makeup, labor production rates, crew mix, equipment requirements, and assumed sequencing of work efforts, plus other support documents, is prepared by cost professionals who are assigned to complete the hazardous waste estimate. These cost professionals define and evaluate assumed/known cost risks utilized in the preparation of a specific hazardous waste cost estimate. The overall probability of a hazardous waste project not exceeding budgetary levels can be equated to the quality and validity of the information contained within the basis of estimate and the evaluation of the cost components used in the preparation of the cost document.

6.9 SUMMATION

The following is a summation of significant cost drivers impacting cost estimates for hazardous waste projects.

6.9.1 Primary Cost Drivers for Hazardous Waste Project Cost Estimates

There are four components whose interrelationships, when combined, are the primary cost drivers for any hazardous waste project. These four elements, in order of cost impact, are:

1. Classification of the hazardous materials within the confines or boundaries of the project site.
2. This classification, in turn, establishes the worker protection level, that is, Level A, B, C, or D, to be used on, and within, the hazardous waste project.
3. Ambient temperatures at the project work site. This component sets the frequency and duration of work breaks during the work day/shift for light, moderate, and heavy work efforts in worker protection Levels A, B, C, or D.
4. Expenditure of kilocalories/hr. This criterion defines the level of work being performed by the worker, for example, light, moderate, or heavy.

6.9.2 Secondary Cost Drivers

Other secondary cost drivers for cost estimates prepared for hazardous waste projects are listed in Section 6.4 of this chapter and are dependent upon the time duration (in weeks, months, and years) of the hazardous waste project.

6.9.3 Other Cost Considerations

One of the other cost considerations that should not be overlooked relates to *possible* relocation of populace surrounding the hazardous waste project. Your motto here is: *remember Love Canal*, which had significant cost repercussions associated with relocating the populace, resettling in costs, purchasing of affected land and structures, etc., plus other legal costs.

6.9.4 Last But Not Least

Preparing cost estimates for any hazardous waste project requires *time*. Do not permit yourself or your staff to be *pushed* or *rushed* into preparing a hazardous waste project cost estimate "on the back of an envelope," nor attempt to complete any cost estimate for these unique projects without reviewing the cost drivers identified and following the approach recommended within this chapter. Following a *structured* approach will save on tranquilizers, and reduce "on-the-job heartburn" for environmental, estimating, cost engineering, and other industry professionals. The structured approach and cost drivers discussed in Chapter 6 are applicable now (1992), and will be applicable 5 and even 50 years from now. This is because calculations for all cost drivers impacting primary, secondary, and other cost components are based on a structured and subjective evaluation of factors by each cost professional for his or her *specific* and *unique* hazardous waste project. Also, it is recommended that practicing cost professionals use their own company's cost data base when preparing hazardous cost estimates.

APPENDIX

Example

The following example is a theoretical application of a labor productivity factor developed in this chapter and used with an existing company cost data base:

Assume your company has received a request for quote (RFQ) for a preliminary cost proposal for the forming, rebar installation, and placement of 300 yds (M3) of concrete for an owner-excavated equipment foundation at an existing hazardous waste site. This hazardous waste project dictates worker production Level B. The owner has identified this segment of the work to be a moderate work effort level. Site temperatures during this operation are expected to be 65°F (18.3°C).

Your company is requested to provide an RFQ for the estimated site work hours, estimated costs for Level B PPE ensembles and the labor and material costs to place the 300 cyd (M3) of concrete. The company estimating data base for the placement of a similar concrete foundation (in a nonhazardous waste environment) is 18 work hours/cyd (M3), with a labor and material unit cost of $570/cyd (M3). This same data base presents a labor and material split of 40% and 60%, respectively.

A recommended estimating approach would be:

1. Using your company's Table 6.2, extract the labor and material multiplier matching the RFQ criteria (moderate work effort, Level B PPE and site temperatures 65°F (18.3°C)) = 1.71.*
2. Using your company's Table 6.2, extract the 100% labor multiplier matching the RFQ criteria = 2.78.*
3. Perform the following calculations

   • Modify the 100% labor multiplier 2.78 to reflect the 40% labor component:

   2.78 × .40 = 1.11

   • Calculate the site work hours for Level B PPE:

   300 cyd (M3) × 18 work hours/cyd (M3) × 1.11 = 6,005

   • Calculate the cost of the Level B ensembles:

   6,005 site work hours (Level B) × 10.00*/crew work hour = $60,050

   • Calculate labor and material costs to place 300 cyd (M3) of concrete:

   $570/cyd (M3) (company data base) × 300 cyd (M3) × 1.71 = $292,410

---

*The labor and material multipliers used in this example were extracted from Table 6.2.

- Summarize and present:

Level B site work hours	6,005
Costs for Level B PPE ensembles	$ 60,050
Place 300 cyd (M3) of concrete	$292,410

RFQ costs (excluding Mob, Oh'd, and Profit) $352,460

7
Contingency Estimating for Environmental Projects

R. F. Shangraw, Jr.
Project Performance Corporation, Sterling, Virginia

7.1 INTRODUCTION

The issue of cost growth has plagued environmental projects for the past two decades. Studies by the General Accounting Office (1992a,b), the Office of Technology Assessment (1991), the University of Tennessee (Russell, Colglazier, and English, 1991) and others (U.S. Army Corps of Engineers, 1990; Kilby, 1992) have documented the incidence of cost growth in both public sector and private sector projects. Consequently, the ability to adequately estimate and manage contingency for environmental projects has become a fundamental component of the cost estimating and control process.

Unfortunately, few analyses have provided adequate methods for addressing cost risks in clean-up projects. The most common approach to eradicating cost growth is to revisit the basis of the estimate and to ensure that the current estimate accounts adequately for known scope as well as uncertainties surrounding the accomplishment of the current scope of work. Some project managers expand upon this process by evaluating the potential changes in scope and incorporating these costs into the estimate to address these risks. Finally, the potential increases in costs over time that are reflected in the cost of basic goods and services underlying the project are included as escalation.

Much of the fundamental theory and practice surrounding contingency estimating for large, capital projects is applicable to environmental remediation projects. However, because of the relatively small amount of actual clean-up experience and some of the peculiarities of the regulatory process, cost risk analysis offers some special challenges for environmental project managers. This chapter will define contingency, discuss the importance of contingency in environmental projects, identify major drivers of cost risk, and provide a review of existing methods for calculating contingency for environmental projects.

7.2 BACKGROUND

In the mid-1980s, several studies were conducted that addressed cost risk and contingency issues for hazardous waste clean-up projects. These first-of-a-kind studies were aimed at identifying the magnitude of cost overruns in clean-up projects as well as noting the causes of cost growth. Most of the early studies were based on the experiences gained through the U.S. EPA's Superfund Program. By the middle to late 1980s, only a handful of sites had been remediated under the Superfund program and these sites served as a basis for most of the early analysis of contingency and cost risk in clean-up projects.

One study, which was conducted by Putnam, Hayes, and Bartlett for the U.S. EPA in 1987, found a substantial problem with cost growth in 18 completed Superfund projects. The Putnam, Hayes, and Bartlett study (1987) compared cost estimates at the Record of Decision (ROD) stage with final costs for Superfund projects and found that the cost growth ranged from 18% to 179% on this small set of projects. In fact, the average underprediction for all sites was 87%.

In 1988, the U.S. General Accounting Office released a report on cost growth in the U.S. EPA Superfund program. Construction cost growth at the 25 sites examined by GAO ranged from −13% to 99%. In particular, the study found that nonroutine clean-up projects experience, on average, over a fourfold increase in cost growth when compared to routine projects. Nonroutine projects included the excavation of contaminated soils, barrels, lagoons, and other hazardous materials. Although the results were more positive than the Putnam, Hayes, and Bartlett study, the study concluded that cost growth may be more severe under the SARA amendments to the Superfund program because more permanent solutions and thus more "nonroutine" construction will be required.

Since these early studies, a number of additional studies have been conducted on the extent and causes of cost growth in environmental projects. Throughout this chapter, the results of these studies will be discussed. In general, the studies reach the same conclusion—that cost growth is a substantial problem in environmental projects.

7.3 WHAT IS CONTINGENCY?

There are numerous definitions of contingency. In fact, lack of consistency in the definition of contingency causes a great deal of confusion in the cost estimating arena. The American Association for Cost Engineering recently published a definition of contingency in its *Skills and Knowledge of Cost Engineering*, (1992):

> . . . an amount added to the estimate to allow for changes that experience shows will likely be required. This may be derived either through statistical analysis of past project costs or by applying experience gained on similar projects. Contingency usually does not include changes in scope or unforeseeable major events such as strikes or earthquakes.

This definition highlights one of the most important dimensions of contingency for environmental projects: the stipulation that contingency does not include changes in scope. One of the most perplexing problems in environmental projects, especially in their early phases, is the accurate identification of the scope of work. Scope definition is crucial because it frames the baseline against which project changes occur, and thereby ultimately provides a basis for allocating additional resources to the project.

Traditionally, costs that are incurred outside of the original scope of the project have not fallen under the purview of contingency funds. In fact, many estimators would argue that they cannot be held accountable for scope changes and therefore will not estimate contingency requirements for these potential changes. Consequently, two types of contingency calculations can be made for any project: (1) contingencies that address cost uncertainties within the current scope of work (*within-scope contingency* or "known knowns"), and (2) contingencies that address cost uncertainties that are due to changes in the scope of work (*out-of-scope contingency* or "known unknowns"). Table 7.1 lists the most significant contributing factors to within-scope and out-of-scope contingencies.

Unfortunately, in the environmental arena, the problems associated with scope definition are much more severe than in traditional construction environments. In part, poor scope definition is a byproduct of the environmental project lifecycle. In the early stages of an environmental remediation project, the scope of remedial construction activities is highly dependent upon the outcome of the remedial investigation. Early estimates

Table 7.1

Factors Contributing to In-Scope Cost Risk	Factors Contributing to Out-of-Scope Cost Risk
-Change in Labor Productivity (e.g., a higher level or worker protection is required)	-Change in Quantities (e.g.,increase in cubic yards of soil to remediate)
-Change in Unit Cost Rate (e.g., higher than expected unit cost for a treatment technology)	-Change in Technology (e.g., decision to incinerate versus steam stripping)
-Poor Site History (e.g., few historical records on the nature of contamination)	-Change in Schedule (e.g., 6 month delay in starting remedial construction due to shortage in funding)
-New Technology (e.g., early implementation of a new treatment technology)	-Natural Disaster (e.g., flood, hurricane, tornado, etc.)
-Remediation at an Active Site (e.g., a corrective measure inside the boundaries of an operating facility)	-Change in Regulations (this factor often leads to changes in remediated quantities or technical approach)
-Poor Site Characterization (e.g., identification of new contaminants after starting detailed design)	-Change in Off-Site Disposal Options (e.g.,, state restricts out-of-state disposal in secure landfills)
-Inappropriate Contract Form (e.g., letting a fixed price contract for a highly uncertain scope of work)	
-Limited Experience of Project Manager	
-Unforeseen Litigation	
-Unforeseen Permitting, Licensing, Bonding, and Insurance Requirements	
-Poor Safety and Health Plan	

for the cost of remediating a site may be required before completely (or even partially) assessing the nature and extent of the contamination at the site. Consequently, quantities and even basic project data are not available for these early estimates. In fact, the norm in most environmental projects in their early stages is for the scope to change dramatically until the Record of Decision or a preliminary engineering estimate is developed for the site. Even with a Record of Decision in hand, there are many documented cases of poor estimation of quantities of material to be remediated and misestimation of other basic project information.

This problem is highlighted in Figure 7.1. This figure shows the actual cost curve of clean-up projects completed over the past decade. Even at

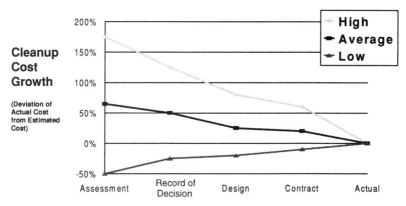

Figure 7.1   Cost curve of projects completed over the last decade. [Adapted from Schroeder and Shangraw (1990b).]

the Record of Decision stage, projects on an average are likely to require 50% contingency, which demonstrates the large amount of risk in these projects. The other important message of this figure is the wide variability in cost growth. Whereas the average needed contingency is relatively high at a more advanced stage of the project, the range around this average percentage is also relatively broad. Again, at the Record of Decision stage, over 66% of the projects (one standard deviation) encounter cost growth that ranges from -25% to $+125\%$.

One solution to the scope definition problem is to explicitly identify and estimate contingencies for within-scope changes and potential out-of-scope changes. This approach has been employed by a number of large clean-up projects. For example, the U.S. Department of Energy's Weldon Spring Site estimates a more traditional contingency for within-scope changes and also computes an *allowance for change (AFC)* for out-of-scope changes.

7.4 THE IMPORTANCE OF CONTINGENCY FOR ENVIRONMENTAL PROJECTS

Given the high degree of cost consciousness among project managers today, the estimation of contingency becomes an inherent source of concern. Furthermore, given that most clean-up projects represent the remediation of past activities and not the production of new goods or services, most companies, as well as the public sector, want to minimize their costs. Unfortunately, there are many sources of cost risk in environmental projects. In almost all cases history has demonstrated the need for additional funding to complete these projects adequately. Furthermore, as many of the easier sites are remediated in the early stages of the large public sector and private sector clean-up programs, many of the more challenging projects must be addressed in the coming years.

A serious source of concern for owners and operators is that contingency will be overestimated and then not properly managed throughout the course of a project. In fact, the estimation of needed contingency must go hand in hand with an adequate system for managing the distribution of contingency funds. Without a system for managing contingency, the likelihood of full expenditure of contingency dollars, whether needed or not, increases substantially. On the other hand, attempts by management to minimize contingency funds leads to a number of other issues. For example, estimates are more likely to be "gold plated," with either additional labor hours or quantities, if contingency is minimized or eliminated from estimates. As with any large project, some flexibility in funding (with proper management control) is necessary to achieve project outcomes.

Most importantly, an understanding of the true costs of an environmental project can better assist decisionmakers in prioritizing and executing these projects. The relationship between cost and human health risk reduction in environmental projects must address contingency or cost risk. Projects with a high likelihood of reducing human health risks but with a high potential for cost overruns must be carefully weighed against projects that have lower potential for cost overruns but with the same potential for reducing human health risks. Although the magnitude of the cost of a project is of primary consideration, the likelihood of cost growth can easily alter funding decisions under different prioritization schemes.

7.5 COST RISK DRIVERS IN ENVIRONMENTAL PROJECTS

There are many sources of cost risk in environmental projects. Although environmental projects will face different types of risk, most of the major cost risk drivers can be categorized into the following four classes:

7.5.1 Scope Definition

Perhaps the largest source of cost risk is simply the degree to which the project is adequately defined. In fact, according to a recent Construction Industry Institute study (1986), poor scope definition is the leading cause of cost overruns in major construction projects. The problems with scope definition in environmental projects have been noted in several studies (Poldy, Shangraw, and Shangraw, 1992). In part, adequate project definition is dependent upon the outcome of the assessment phase (CERCLA remedial investigations [RI] and RCRA facility investigations [RFI]). Also, project execution factors (laboratory availability, labor availability, equipment availability, etc.) are often assumed for environmental projects, which can contribute substantially to cost growth.

7.5.2 Contaminant and Contaminated Media Complexity

Complexity is a very ambiguous term. Nevertheless, more complicated, intricate, multifaceted projects offer more opportunities for cost misestimation and therefore tend to engender more cost risk. The relationship between project complexity and cost growth has been documented for environmental projects (Schroeder and Shangraw, 1990a and 1990b), and for projects in the oil and chemical industry (Merrow, 1981). In cleanup projects, complexity has a number of dimensions. Generally, projects with multiple classes of contaminants present (e.g., volatile organics and heavy metals) and with different types of contaminated media (e.g., groundwater contamination and contaminated unsaturated soils) face a greater chance of incurring cost growth.

7.5.3 Technical Sophistication

Cleanup projects that employ innovative or first-of-a-kind technologies are more likely to need additional contingency funds in the short term. In the long term, however, cost risk should decrease as utilization of the new technology improves. A study performed at Oak Ridge National Laboratory documented such a case, and also provides estimates for cost growth at National Priorities List (NPL) projects with varying degrees of technical complexity and intensity (Doty, Crotwell, and Travis, 1991).

7.5.4 Regulatory Uncertainty

Regulatory adjustments and new legislation and interpretations often lead to changes in scope of projects. Projects are affected by existing regulations, or may be targets of new policy actions. The direction of cost risk caused by regulatory-driven scope changes is usually upwards. In most cases, the need for regulatory-driven contingency is driven by a change in project scope. In some cases, the change in regulatory driver may be expected and contingency can be evaluated deterministically based on this potential change. This new contingency would be determined by costing the scope of the project as driven by the expected change. Differing scenarios contingent on which legal status prevails could be postulated, with risk determined by the probability of occurrence of the scenarios.

One common misconception is that the same factors that drive the cost of a project also contribute to cost risk. This is especially true for quantities of waste to be remediated. In fact, larger projects do not incur greater cost risk (on a percentage basis) than do smaller projects.

7.6 CONTINGENCY ESTIMATING METHODOLOGIES

There are numerous methods for calculating needed contingency for an environmental project. The following discussion has been divided into two main categories. The first section describes contingency estimating at the project level. Here the objective is to develop an estimate of cost risk by examining the project as a whole. The second section examines contingency estimating techniques that estimate cost risk at a lower level in the project (e.g., at the activity or task level) and then aggregate these sub-project level estimates to obtain needed contingency at the project level.

7.6.1 Project-Level Contingency Estimating

Needed contingency for environmental projects is often calculated on a project basis. Projects that are in the early phases of development are rarely defined in sufficient detail to conduct a contingency analysis at the activity level. In the next section of this chapter, several of these detailed approaches are described. Another reason for estimating contingency at the project level is to address risks and uncertainties that are not easily equated with a specific line item in the estimate.

Project Definition Rating

A number of project definition rating or scoring systems have been developed for environmental projects to estimate needed contingency. The pioneering work in this area was contributed by John Hackney (Hackney, 1989). Users of these scales score a project on a variety of factors. The score represents the degree to which the factor can influence the project. The factors are then weighted, summed, and categorized into needed contingency percentages. Table 7.2 summarizes the major factors from Hackney's project definition rating system. The most useful scoring systems are calibrated with actual data. Hackney, for example, calculated rating scores for 15 hazardous waste projects, examined actual needed contingency for these projects, and then established statistical ranges of needed contingency based on different rating scores.

Table 7.3 shows a project definition rating system used by Dames and Moore's Albuquerque office for large environmental projects. This system is based on over a decade of cost estimating experience with hazardous and radioactive waste remediation projects.

Risk and Uncertainty Analysis

Another approach to estimating needed contingency at the project level is to conduct a risk and uncertainty analysis. This analysis is highly par-

Table 7.2 Major Factors Underlying Hackney's Hazardous Waste Project Definition Rating

FACTOR	MAXIMUM SCORE
General Project Basis	1130
Hazardous Material Definition	240
Location Data	300
Objectives	190
Remedial Method Background	200
Disposal Conditions	200
Remedial Design Status	675
Remedial Plans and/or Flow Sheets	130
Monitoring Systems	100
Support Requirements	70
Equipment Selection	155
Management Plans	70
Review of Remedial Design	150
Engineering Design Status	358
Layouts	60
Line Diagrams	50
Auxiliary Equipment, Type and Size	50
Buildings, Type and Size	40
Yard Improvements, Type and Size	58
Review of Engineering Design	100
Detailed Design	96
Drawings and Bills of Materials	46
Drawing Reviews	50
Field Performance Status	50

Adapted from Hackney (1989).

Table 7.3 Sample Contingency Analysis for Line Item Project Estimates

	Relative Weight 0 to 10: 0 = N.A., 10 = Important	Probability Score 1 to 10: 0 = Low Risk, 10 = High Risk	Weighted Score
Engineering Design & Inspection Design Completeness Site Selection/Conditions Design Complexity Design Schedule Life Safety/Security	0	0	0
Construction Types of Construction Construction Complexity Design Completeness Market Conditions Method of Accomplishment	0	0	0
Special Facilities/Equipment Requestor's Specifications Maturity of Technology Design Completeness Method of Accomplishment	3 3 10 5 5	5 5 8 4 4	15 15 80 20 20
Standard Equipment Specification Completeness Quantity Accuracy Price Accuracy Method of Accomplishment			
Project Management Complexity Schedule Quality Assurance/Control			
Total Weighted Score (Total possible score = 260)	26	26	**150**

Score Ratio (Total weighted Score / 260): 150/260 = **0.58**

NORMAL MAXIMUM CONTINGENCY:
Planning / Feasibility	X 35%	Title II	___15%
Budget / Conceptual	___25%	Portion Above Low Bid	___10%
Title I	___20%		

BASE CONTINGENCY:
    Score Ratio x Normal Maximum Contingency = 0.58 x 35% = 20.30%

TYPES OF CONSTRUCTION:
    Quality Assurance Level I, II, or III: Add 0 to 10%   = 5.00%
    Underground Lines Not Clearly Defined: Add 2 to 5%   = 0.00%
    Contaminated Areas (describe): Add 5 to 10%   = 0.00%

Other special requirements, construction/design uncertainty; add %   =2.00%

Total Item Task Contingency   27.30%

ticipatory, requiring input from all members of the project team. The project team first identifies the major sources of risk and uncertainty for the project. The list should be limited to the top 20 sources and include factors that contribute both to in-scope and out-of-scope changes. Then, after discussion among the team, the likelihood of encountering each risk source is estimated. To simplify the analysis, these probabilities can be lumped into three categories: highly likely (e.g., greater than 90% probability of occurrence); likely (e.g., between 50% and 90% probability of occurrence); and somewhat likely (e.g., less than 50% probability of occurrence). In addition to the probability assessment, the potential cost (and/or schedule) impact of the risk source is roughly calculated. This data is then evaluated by the project team. Table 7.4 displays a sample risk and uncertainty analysis for DOE's Grand Junction Program. Unfortunately, the system breaks down when multiple risk sources are interdependent and have varying conditional probabilities.

Historical Analysis of Comparable Projects

A recently completed study conducted by Independent Project Analysis, Inc. examined the factors that contribute to cost and schedule overruns in large environmental projects. The study used a database of 184 completed remediation projects and developed a statistical relationship between cost growth and a variety of descriptive project parameters. Because the database includes Superfund, Department of Defense, private sector projects, and DOE projects, the results are more representative of the industry as a whole. In short, the study established a number of statistical relationships between cost growth during the assessment and cleanup phase of remediation projects and certain project characteristics. These project characteristics included the complexity of the contaminated media, the nature of the contamination, the degree of project definition at the time the estimate is developed, the complexity associated with the geology and subsurface features at the site, and a number of other factors. The results of the study have been summarized in a number of conference presentations (Schroeder and Shangraw, 1990a and b; Schroeder, 1990). The statistical models derived from this study can be used in a predictive fashion to estimate needed contingency for projects.

7.6.2 Task-Level Contingency Estimating

Many project managers prefer to examine contingency at the major task level. For a CERCLA RI/FS project structured in a traditional manner,

Table 7.4 Sample Risk Analysis to Derive Site Budget Contingency Requirements ($M)

ITEM	IMPACT ($M)	PROB	EXPECTED VALUE	<50% RISK TOTAL	> 50% <90% RISK TOTAL	>90% RISK TOTAL	>50% RISK PER FISCAL YEAR					
							FY92	FY93	FY94	FY95	FY96	FY97
1. NEW BURROW FOR EROSION PROTECT	$1.90	20%	$0.38	$1.90								
2. ADD'L NRT QUANTITIES	$2.75	60%	$1.65		$2.75		$1.00	$1.75				
3. RESTORE COTTER	$0.25	40%	$0.10	$0.25								
4. RESTORE HAUL ROAD	$0.25	75%	$0.19		$0.25				$0.25			
5. SITE EXT; VP NOT COMPLETED 8/93	$2.30	40%	$0.92	$2.30								
6. OSHA 40 HR TRAINING	$0.75	50%	$0.38		$0.75		$0.38	$0.37				
7. METHOD USED FOR DECON TRUCKS	$0.50	50%	$0.25		$0.50		$0.50					
8. OVERSIZED RUBBLE IN PILE	$0.40	20%	$0.08	$0.40								
9. ADD'L CHARGE FOR HAL, DOT REQT	$10.00	50%	$5.00		$10.00		$5.30	$4.70				
10. SHIFT IN COST		40%										
11. PROTECTIVE CLOTHING	$0.15	80%	$0.12		$0.15		$0.06	$0.09				
12. CONTINGENCY FOR VP REFUSALS												
13. POST CELL CLOSURE	$1.00	85%	$0.85		$1.00							$1.00
14. ADD'L CONTAMINATED MATERIAL	$3.20	10%	$0.32	$3.20								
15												
16												
17												
18												
19												
20												
SITE TOTAL	$23.45		$10.23	$8.05	$15.40		$7.24	$6.91	$0.25			$1.00

this would mean identifying cost uncertainties for the remedial investigation work plan, the remedial investigation, the feasibility study, treatability studies, and any other major assessment task. Task level contingency estimating forces the project team to study and estimate cost uncertainties at a more detailed level. Unfortunately, this approach also leads to the possibility of overlooking larger drivers of cost risk. For example, a regulatory change may pervade the entire estimate and therefore may not be accounted for in the contingency for a specific task. There are several methods for calculating needed contingency at the task level.

Range Estimating

Range estimating has been applied to many projects as a means of estimating needed contingency at the task level. This approach has been documented for a variety of projects (Hudson, 1992; Curran, 1989). To estimate the project contingency, an item-by-item analysis of the scope of work should be completed. At a minimum, this analysis should occur at the work element level and, preferably, at the next lower level in the WBS. In addition, this analysis should be completed for each estimate type (planning, conceptual, and detailed). The project team should then develop a list of the major cost drivers. These categories effectively establish the degree of definition surrounding each WBS element and might include:

- Regulatory/permits
- Characterization data
- Design criteria
- Design detail
- Project management
- Labor/productivity
- Quantities
- Work stoppages
- Procurement strategy

For each item in the WBS and for each type of estimate, the project team should quantify the high, low, and most likely cost impact of each cost driver category. These costs are then summed across the cost driver categories and across the WBS elements to develop a high, low, and most likely cost impact scenario. The project team also should adjust its analysis to address any significant interactions among the cost drivers and the WBS categories. A recent paper documents a method for developing uncertainty models for environmental restoration projects (Hudson, 1992). This analysis should be updated on a routine basis and should be subjected to a comprehensive review by management.

Table 7.5   Contingency Percentages for Selected DOE Installations

TASK	Mound	Pantex	Sandia-AL
RI or RFI Workplan	10%	0%	20%
RI or RFI Field Work	50%	40%	50%
RI or RFI Report	20%	10%	50%
CMS or FS	20%	30%	50%
CMI Program Plan or ROD	20%	10%	20%

Standard Values for Specific Tasks

Another approach is to use specified ranges of needed contingency, based
on historical analysis, for specific tasks. Table 7.5 provides a summary
of specific contingency values used by a sample of U.S. DOE sites in addressing cost risk in assessment projects. Figure 7.2 provides a similar
set of ranges for remediation projects based on analysis of needed contingency for remedial design and construction projects. Table 7.6 lists

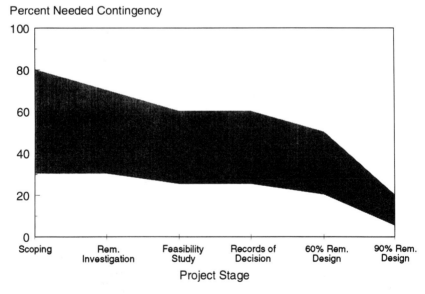

Figure 7.2   Contingency ranges for remedial construction projects. [Adapted
from U.S. Department of Energy (1991).]

Table 7.6 Recommended Contingency Percentages for Decontamination and Decommissioning Projects

Category	Contingency Percent
Engineering Project Management, Demolition Management	15%
Utility and DOC Staff Costs	15%
Decontamination	50%
Contaminated Component Removal, Contaminated Concrete Removal	25%
Steam Generator, Pressurizer, PWR Reactor Coolant Pumps & Piping Removals, BWR Recirculation System Pumps and Piping Removals	25%
Reactor Vessel and Internals Removal	75%
Reactor Packaging	25%
Reactor Shipping	25%
Reactor Burial	50%
LSA Packaging	10%
LSA Shipping	15%
LSA Burial	25%
Clean Component and Concrete Removals, Clean Waste Disposal	15%
Supplies and Consumables	25%

Adapted from Atomic Industrial Forum (1986).

recommended contingency percentages for major tasks for decontamination and decommissioning projects. Although this approach lacks the
analysis present in several of the preceding techniques, it provides a simple framework for estimating contingency with a shortage of estimating
resources.

7.7 CONCLUSION

In an attempt to ultimately improve the estimation and management of
contingency, this chapter provides a survey of current techniques for esti-

mating contingency for environmental remediation projects. This chapter identifies a number of approaches for estimating contingency in the environmental restoration area and highlights the major areas of concern for estimating contingency for this class of projects. In the dynamic environment in which environmental programs operate, the estimation and effective management of contingency plays an important role in the ultimate achievement of an organization's objectives. Without a comprehensive program for addressing cost risk in these projects, the problems with cost growth are likely to be exacerbated.

Much of the new research conducted by EPA, DOE, DoD and private sector organizations in this area will provide a strong foundation for future contingency estimating practices. Furthermore, this research will also lead to a greater appreciation of the uncertainties surrounding environmental restoration projects, as well as the complexities inherent in this type of estimating process. The challenge facing environmental programs over the next several years is to incorporate the results of the contingency research, where appropriate, into field and contractor estimating practices. This integration process will require a strong leadership role on the part of management in the areas of cost and schedule estimating guidance, training, and education. Fortunately, as a relatively new area of concern, environmental programs have an opportunity to establish procedures for estimating and managing contingency, as well as processes for tracking and evaluating cost growth in these programs.

ACKNOWLEDGMENT

I would like to thank Anton Quist for his assistance with this chapter.

REFERENCES

American Association of Cost Engineers, Inc. (1992). Skills and Knowledge of Cost Engineering, 3rd ed, Morgantown, WV.

Atomic Industrial Forum (1986). Guidelines for Producing Commercial Nuclear Power Plant Decommissioning Cost Estimates, AIF/NESP-036, Bethesda, MD.

Burgher, B., M. Culpepper, and W. Zeiger, (1986). Remedial Action Costing Procedures Manual. Prepared for the United States EPA; EPA/600/8-87/049.

Doty, C. B., A. G. Crotwell, and C. C. Travis, (1991). Cost Growth for Treatment Technologies at NPL Sites, Oak Ridge National Laboratory; ORNL/TM-11849.

Construction Industry Institute (1986). Scope Definition and Control. Bureau of Engineering Research, The University of Texas at Austin, TX.

Curran, M. W. (1989). Range Estimating, *Cost Engineering*, 31.

Hackney, J. W. (1989). Accuracy of Hazardous Waste Project Estimates. *1989 Transactions of the American Association of Cost Engineers*, American Association of Cost Engineers, Morgantown, WV.

Hudson, C. (1992). A Technique for Assessing Cost Uncertainty, *1992 Transactions of the American Association of Cost Engineers*, American Association of Cost Engineers, Morgantown, WV.

Kilby, J. L. (1992). Lessons Learned from Remedial Design/Remedial Action, *Hazardous Materials Control*, 5.

Merrow, E. W. (1981). Understanding Cost Growth and Performance Shortfalls in Pioneer Process Plants, The Rand Corporation, Santa Monica, CA.

Poldy, J. L., W. R. Shangraw, and R. F. Shangraw, Jr., (1992). A Project Definition Methodology For Environmental Restoration Projects. *1992 Transactions of The American Association of Cost Engineers*, American Association of Cost Engineers, Morgantown, WV.

Putnam, Hayes, and Bartlett (1987). Preliminary Results Regarding Cost Escalation at Superfund Sites, Prepared for U.S. Environmental Protection Agency, Washington, D.C.

Russell, M., E. W. Colglazier, and M. R. English, (1991). Hazardous Waste Remediation: The Task Ahead, Waste Management Research and Education Institute, University of Tennessee, Knoxville.

Schroeder, B. R., and R. F. Shangraw, Jr., (1990a). Understanding Cost Drivers for Remedial Investigations/Feasibility Studies. Proceedings of the 7th National Conference on Hazardous Waste and Hazardous Materials, St. Louis, MO.

Schroeder, B. R., and R. F. Shangraw, Jr., (1990b). Parametric Tools for Hazardous Waste Projects; *1990 Transactions of the American Association of Cost Engineers*, American Association of Cost Engineers, Morgantown, WV.

Schroeder, B. R. (1990). Cost Inaccuracies in Superfund Projects: Strategies for Building Better Estimates, Proceedings of the 11th National Superfund Conference, Washington, D.C.

United States Army Corps of Engineers Water Resources Support Center (1990). Hazardous and Toxic Waste (HTW) Contracting Problems—A Study of the Contracting Problems Related to Surety Bonding in the HTW Cleanup Program; IWR REPORT 90-R-1.

United States Congress General Accounting Office (1992a). Nuclear Health and Safety: More Can Be Done to Better Control Environmental Restoration Costs, GAO/RCED-92071, Washington, D.C.

United States Congress General Accounting Office (1992b). Superfund: EPA Cost Estimates Are Not Reliable or Timely, GAO/AFMD-92-40, Washington, D.C.

United States Congress General Accounting Office (1992b). Superfund: EPA Cost Estimates Are Not Reliable or Timely, GAO/AFMD-92-40, Washington, D.C.

United States Congress General Accounting Office Report to Congressional Requestors (1988). Superfund: Cost Growth on Remedial Construction Activities, GAO/RCED-88-69, Washington, D.C.

United States Department of Energy (1991). EM-40 Baseline Workshop Manual, Office of Environmental Restoration, Washington, D.C.

United States Congress Office of Technology Assessment (1991). Complex Cleanup: The Environmental Legacy of Nuclear Weapons Production, OTA-O-484, Washington, D.C.

8
Decisions in the Marketplace

Kenneth R. Cressman
Jacobs Engineering Group, Denver, Colorado

Bruce A. Martin
Black & Veatch, Overland Park, Kansas

8.1 INTRODUCTION

In the late 1970s and early 1980s many companies embarked on a strategy of diversification for survival. Both petroleum resource-based companies and nonpetroleum resource-based companies (such as engineering and construction firms) acquired minerals ventures in attempts to expand and bolster their profit base and to solidify leadership positions in response to declining oil prices and uncertain supply. At that time, with most mineral prices hovering near their peak, entry into these markets certainly appeared profitable and compatible with these companies' current operations. However, economic stagnation, localized political upheaval in developing countries, and the age/decay of some of the mining operations have made this strategy of diversification rather unattractive in the past few years. Figure 8.1 highlights this "rise and fall of business," which can be cyclical in nature, that is, when earnings increase, debt decreases, and vice versa. Lower commodity priuces left these companies with difficult choices between several less-than-optimum scenarios. As shown in Figure 8.2, these companies must first decide whether to attempt to maintain the business or to close it altogether. This is a difficult decision to make in a seemingly *Catch-22* situation: market prices are down, so the business must close because it cannot continue operating at a loss; there is a potential buyer, but a return on shutdown costs can not be realized

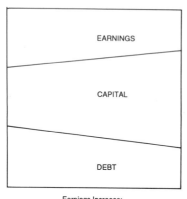

Figure 8.1 The rise and fall of business.

Shut-Down Costs

- Decommissioning
- Mothballing
  - Deactivation of Systems
  - Winterization
- Personnel
  - Early Retirement
  - Lay-off Compensations
- Security Precautions
- Disposal of Inventories
- Taxation Credits

Holding Costs

- Maintenance & Upkeep
- Utilities
- Insurances
- Taxes
- Support Services
  - Security
  - Bookkeeping

Re-Start of Operations

- Demothballing
  - Clean-up
  - Physical Inspection
  - Reconditioning & Servicing
- Recommissioning
  - Refurbishment/Replacement
  - Energizing Systems
- Recall of Personnel
  - Possible New-Hires
- Orientation & Retraining
  - Operations
  - Safety
- Restock Inventories
  - Working Capital
  - Supplies

Figure 8.2 Temporary closure.

unless modernization occurs. If modernization occurs, will it put the business in a competitive position to continue? This chapter will define these options and attempt to highlight some of the problems that these companies must address and solve.

8.1.1 Maintenance of Business

Certainly one option that these companies must consider is to maintain the business. When the company desires to continue operating, this often means the implementation of new technology, upgrading or refurbishment of equipment, and reduction in operating costs, including renegotiation of union labor contracts and reduction in work force and support services. Companies must become "lean and mean" in an attempt to stay competitive. One means of achieving this is to procure and refurbish used processing equipment. Virtually new equipment could be procured for as little as 30 cents on the dollar in the mid-1980s. In isolated cases, equipment trades of like value through a broker are possible, with minimal fees imposed. Other possible options are the forming of a joint venture to run the operation or simply maintaining the holdings for speculative "better times." Maintaining the holdings requires cash outlay for the payment of taxes, insurance, shutdown support, and loss of revenue. In our society government subsidies are not acceptable alternatives as is the case in some other parts of the world. The capitalistic viewpoint has always maintained that initiative, ingenuity, resourcefulness, and determination will succeed; this expresses the "survival of the fittest" idealogy to its fullest. However, it is the opinion of the authors that these companies must now face up to the poor management decisions of the past, including the failure to reinvest into profit-generating facilities during profitable times to ensure desirable rates of return for the future. One alternative that the majority of businesses appears to be following is that of reorganizing management or restructuring business emphasis. Mr. Akio Morita, chairman of the Sony Corporation, is quoted in the November 17, 1986 issue of *U.S. News and World Report*:

> Right now in this country (America), service industries are booming. But if America goes to services and forgets production industries, you must not complain about balance of trade, because you are not producing. Producing real things well appears to be unimportant to many U.S. business executives. They are moving to the "money game"—buy a company, sell a company, take over a company. If you continue this, American industry will completely deteriorate.

If American industry decides to rebuild its competitiveness, it's easy. You already have a base.

Peter Drucker, professor of Social Sciences at Claremont Graduate School, makes the following points in a February 2, 1987 article in *U.S. News and World Report*:

Manufacturing industries are no longer central . . .

The mills that lose the most money are the modernized ones. That's always true of an obsolete industry . . .

The organization of the future will be one that will have far fewer managers and far more specialists with knowledge.

Should a company opt to cease operations temporarily in an effort to retrench, re-evaluate, and restructure, consideration must be given to the costs associated with such a move (Figure 8.3). It is obviously not something to be taken lightly and will most often occur as a long-term commitment. The end result may lead to a permanent closure. At stake more than immediate markets are the concerns of fewer good jobs, shrinking incomes, a declining standard of living, and the threat upon national security ("Will the U.S. Stay Number One?," *U.S. News and World Report*, Feb. 2, 1987, pp. 18–22).

The company that makes the decision to maintain its business or to temporarily halt operations must fully understand the consequences of that decision. It should keep in mind the words of Jonathan Aronson, international relations expert at the University of Southern California: "They tried to do too many things at once. American companies that have focused on a particular niche have been more successful."

The decisions facing these industries are not simple or straightforward by any stretch of the imagination. Each has far-reaching implications, ranging from local economy, competitive position, both in the U.S. and internationally, and liability issues to impacts upon the environment. These businesses can take some solace in the fact that many of their dilemmas are not new. The problems have been around for some time and surface in economic cycles (References, "What To Do With That Old Plant," by Roger Williams, Jr., in December 1953 issue of *Chemical Engineering*.)

While the dilemmas may not be new, the possible solutions to them certainly are. America's businesses must take advantage of new and ever-improving technological advances to assist them. Only by making the hard decisions now can America be prepared for the future.

8.1.2 Facility Closure and Reclamation Costs

For those companies who have decided to cease their ventures, several options must be considered. Of principal concern is the environmental

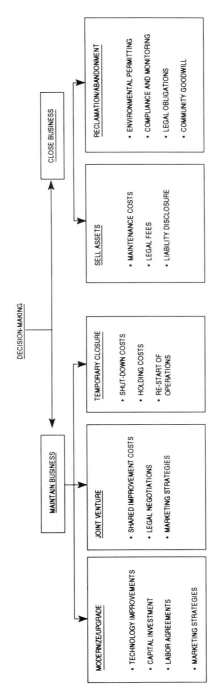

Figure 8.3 Factors to consider when operating at a loss.

liability associated with the property. If a buyer cannot be located to take over the facility, and subsequently any liability, monies will need to be spent on both site cleanup and reclamation. Secondly, if the reserve assets are substantial, the property may be considered a valuable investment by some other developer; however, the present condition may be such that capital expenditure will be required to upgrade the process facilities in order to make the property more attractive to a potential buyer. This is not unlike a home owner spending money for repair and maintenance as a condition of sale on a residence. Reserves may need to be set aside for (future) shared costs of joint environmental liability.

It appears as though there will be increasing future demand for companies engaged in the business of returning closed plant sites to an original and safe condition. Both local communities and state and federal governments are enforcing their concerns of groundwater contamination, air pollution, and hazardous toxins and gases escaping into the environment. This is evidenced by the promulgation of such measures as the Uranium Mill Tailings Radiation Control Act of 1978 (Public Law 95-604, 42 USC 7901), the Comprehensive Environmental Response Compensation and Liability Act (CERCLA) and the Resource Conservation and Recovery Act (RCRA). As steel mills shut down and mines deplete their reserves, regulatory agencies are stepping in to ensure that the companies adequately and responsibly address their liabilities. Throughout the western States, cleanup is under way: mine sites in Montana, New Mexico, and Arizona; tailings in Utah, Colorado, New Mexico, and Nevada; and smelters in various locations are currently undergoing cleanup.

The remainder of this chapter will provide Cost Estimating Relationships (CERs) and unit costs for many of the activities associated with the environmental restoration of a site. The CERs were developed by entering both work hours and dollars into a statistical computer program. The costs provided are based on actual experience and are intended to be used for budget or check level estimates. All costs are in mid-1991 dollars. Many of the costs provided were developed for the Office of Programs/ Project Management and Control of the U.S. Department of Energy (DOE). These costs are currently being used at several DOE sites to prepare estimates for environmental restoration.

8.2 COST ESTIMATING RELATIONSHIPS AND UNIT COSTS

8.2.1 Remedial Investigation/Feasibility Studies (RI/FS)

The first step in conducting any type of environmental restoration at a site is the remedial investigation/feasibility study (RI/FS) process. The

RI is conducted concurrently with the FS, and emphasizes data collection and site characterization. The data collected during the RI support the analysis and decision-making activities of the FS, as well as remedial alternative evaluations through treatability investigations. Figure 8.4 shows a flow diagram of the phased RI/FS process and highlights this interdependency. A cost summary for the overall RI process is shown in Table 8.1. The CER for the overall RI process was:

$$\text{Hours per hectare} = (498 \div \text{hectares}) + 6.5 \tag{1}$$

The overall correlation coefficient for this equation was 0.936. NOTE: Multiply acres by 0.405 to convert to hectares.

The source of the costs shown for the RI process in the succeeding sections was work hour estimates developed by IT Corporation (IT) for remedial investigations at DOE's Rocky Flats plant in Colorado. The reader should apply established location factor adjustments to the CERs as required. The areas covered by the RIs included retention ponds, trenches, process tanks, contaminated soil, surface spill areas, and effluent discharges.

Scoping Process

The first step of the RI is to scope the nature of the contamination problems by collecting and evaluating existing data on the site. The existing information may be found in DOE, U.S. Environmental Protection Agency

Table 8.1 Cost Summary of RI Process

Remedial Investigation Activity	Average Cost per Hectare	Percent of Total
Scoping Process	$ 151	1
Sampling/Analysis Plan Development	626	3
Site Characterization	10,434	54
Treatability Investigation[a]	4,248	22
Data Analysis and RI Report	885	5
Data Management (10%)[b]	1,634	8
Health/Safety Planning	159	1
Community Relations (7%)[b]	1,145	6
Total	$19,282	100

[a] Assumes a Bench Test only and a capital cost of $500,000 for the selected alternative.

[b] Data management and community relations are calculated as percentages of the scoping process, sampling/analysis plan development, site characterization, treatability investigation and data analysis and RI report activities.

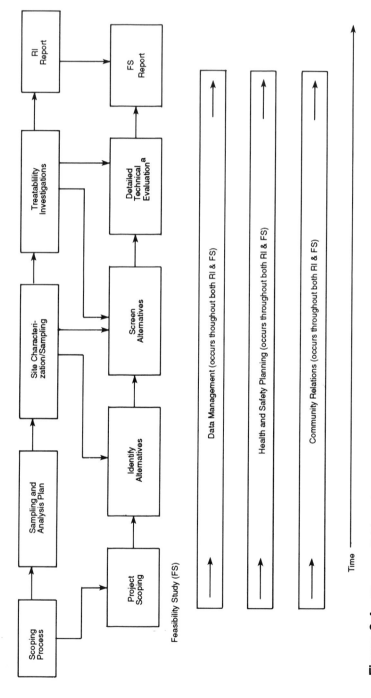

Figure 8.4 The remedial investigation (RI) and the feasibility study (FS). A treatability investigation may also take place during the detailed technical evaluation.

(EPA), site investigation contractor and state (if applicable) files. The existing data are used to compile a site description and chronology of significant events, which are used to evaluate the potential effects and impacts of hazardous substances on public health, welfare, and the environment. The data are also used to determine the need for immediate removal, planned removal, or initial remedial measures (IRMs). These are actions that may be taken at the site prior to the selection of a final remedial action and are intended to protect the public health and environment throughout the remainder of the remedial action process.

The final step of the scoping process is to identify and develop general response actions or categories of remedial actions that will address the contaminants. Costs for the scoping process cover the labor to collect and evaluate the existing data on the site, as well as to compile the site description and to identify the general response actions. The work hours for the scoping process ranged from 1.5 to 20.9 per hectare, with an average of 2.9 work hours per hectare. Assuming a labor rate of $50 per work hour would yield a cost range of $75 to $1,045 per hectare, and an average cost of $144. (NOTE: $50 per workhour was assumed as an average labor rate for environmental professionals. The reader should replace this rate with an actual labor rate based on his/her own company.) Costs for travel, reproduction, and the purchase of maps and other documents should be added to these ranges as appropriate.

The CER for the scoping process was:

$$\text{Hours per hectare} = (26.6 \div \text{hectares}) - 1.5 \tag{2}$$

The overall correlation coefficient for this equation was 0.981.

Sampling and Analysis Plan Development

The second step of the remedial investigation process is the development of a sampling and analysis plan (SAP). The SAP defines the specific field activities required and clearly delineates the number of each sample type, describes the collection methods to be used, specifies each sampling location, and provides an explanation for the selection of the locations. The sample type and collection methods will be determined based on the waste constituents that are known or thought to be present at the site. The SAP will define the sampling locations through the use of a grid system (normally on a 9-m (30-foot) basis for radiological surveying) and will also define the sampling frequency.

Preparation for sampling includes procuring qualified analytical laboratory services, sample containers, and necessary equipment such as protective clothing, safety equipment, labels, chain-of-custody forms, and cleaning, presentation, and packing materials. An operational plan is also developed to specify the responsibilities of the sampling team members, as well as to list procedures for recordkeeping and documentation, equipment use, sampling order, and decontamination. The final step of the SAP development is to estimate the costs for the sampling effort based on the items and activities defined by the SAP.

Costs for SAP development cover the labor to define the number and types of samples required, to specify the sampling locations on the grid system, to define the frequency of sampling, to prepare for the sampling, and to develop the cost estimate to implement the sampling plan. The work hours for SAP development ranged from 5 to 48 per hectare, with an average of 11.9 work hours per hectare. Again, assuming a rate of $50 per work hour, the costs for SAP development ranged from $243 to $2,410 per hectare, with an average cost of $597 per hectare. Costs for travel, reproduction, and the purchase of maps and other documents should be added to these numbers as appropriate.

The CER for SAP development was:

$$\text{Hours per hectare} = (52.3 \div \text{hectares}) + 13.4 \tag{3}$$

The overall correlation coefficient for this equation was 0.880.

Data Management

Data management during the RI involves following specific protocols and guidelines to ensure the validity of the data for decisions made during the FS.

There are two types of information that must be maintained and documented. The first type covers the technical data required for or generated during the remedial investigation, including field samples, log books, and laboratory or engineering analysis data. Chain-of-custody procedures must be followed to document sample possession because the samples are often used as legal evidence. The second type of data is that required to properly control the quality, and to manage and report on the status of the project. This data includes a comprehensive, well-documented Quality Assurance/Quality Control (QA/QC) plan, as well as technical and financial progress reports.

The level of effort required for data management during the remedial investigation is difficult to estimate because it is based on project-specific conditions; however, a rule of thumb of 5% to 10% of the total RI costs may be used.

Health & Safety Plans

A written, site-specific health and safety (H&S) plan will be developed to address information on the site, hazard evaluation, personnel and site monitoring requirements, required levels of personal protection, work limitations, authorized personnel, decontamination, and emergency procedures. Procedures for medical surveillance, recordkeeping, training, and equipment use and maintenance are also defined.

The costs for H&S planning cover the labor to develop the site-specific H&S plan. The range of work hours for this activity was 2 to 4 work hours per hectare. Assuming a rate of $50 per work hour, the range of costs was $101 to $202 per hectare, with an average cost of $152 per hectare. Travel and other costs should be added as appropriate.

Community Relations

Community relations activities during the RI phase include the development of a site-specific community relations plan (CRP), which explains how the impacted community will be informed about activities at the site and how the community will provide input for response decisions. The CRP specifies when progress reports will be provided to the community, the frequency of formal and informal meetings, as well as briefings of local and state officials. Development and implementation of the CRP requires close coordination and communication with local citizens and officials, the EPA, federal agencies (Department of the Interior, U.S. Army Corps of Engineers, U.S. Geologic Survey, and others), and the state.

Although it is difficult to estimate the level of effort required for community relations, a rule of thumb is that 5% to 7% of the total RI costs may be used.

Site Characterization (Sampling)

Site characterization provides data to support the decisions made in the feasibility study, as well as providing information concerning the environmental setting, hazardous substances, environmental concentrations, potential impact on humans, animals, and plants in the area, and the effectiveness of potential remedial actions. Because each site is different, it is difficult to standardize which technical investigations must be performed; however, in general, the investigations that may be required are geological, groundwater, surface water, soil, air, and biological sampling. The samples collected during the site characterization are sent to an analytical laboratory to determine the type and concentrations of the hazardous substances, as well as to assist in determining the severity of the hazards.

The costs to perform the site characterization will comprise the majority of the total cost for the RI. Because the determination of the types and numbers of samples to be collected and analyzed depends on the specific site, it is difficult to provide a "standard" cost for this phase; however, the range of work hours for these activities was 1.2 to 212 work hours per hectare, with an average of 30.8 work hours per hectare. Again, assuming a rate of $50 per work hour would result in a range of costs of $61 to $10,712 per hectare, and an average cost of $1,550 per hectare. The range of costs for other direct costs (ODCs), including subcontractor costs and equipment rental was $69 to $28,131 per hectare. The average cost for ODCs was $2,106 per hectare. The range of costs for sample analysis was $2,361 to $30,891 per hectare, with an average cost of $6,708 per hectare.

The CERs for site characterization were as follows:

$$\text{Work hours per hectare} = (277 \div \text{hectares}) - 4.3 \tag{4}$$

The overall correlation coefficient for this equation was 0.857.

$$\text{ODCs dollars per hectare} = (\$37,428 \div \text{hectares}) - \$6,873 \tag{5}$$

The overall correlation coefficient for this equation was 0.791.

$$\begin{aligned}\text{Sample analysis dollars per hectare} = \\ (\$35,561 \div \text{hectares}) + \$5,520\end{aligned} \tag{6}$$

The overall correlation coefficient for this equation was 0.910.

NOTE: The equations for work hours per hectare and ODCs dollars per hectare reflect low correlation coefficients because of the wide range of sampling activities that could be required for a site. These equations were included as the "best fit" of the data available. Table 8.2 provides unit costs for characterization activities. Table 8.3 provides estimated purchase costs for sampling equipment.

Treatability Investigations

For some sites it may be necessary to perform treatability investigations (bench and pilot studies) to obtain enough data to select and implement a remedial action alternative. Under these circumstances, the bench and pilot studies are performed as part of the RI. Bench and pilot studies performed to provide data for the design and construction of the selected remedial technology are outside the scope of the RI.

Table 8.2   Estimated Unit Costs for Characterization Activities

A. Sampling

   1. Subsurface Soil

Drilling	$110 per vertical linear foot (VLF) (subcontractor cost - includes all Mob, Demob, Overhead, G&A, and Profit)
Technical Labor (includes Geologist, Technician, H&S Watch, QA, and Project Management)	20 workhours per sample, 2 workhours per VLF
	20 workhours x $50/workhour (includes Overhead, G&A, and Fee)
	= $1,000 per Sample = $100 per VLF

   2. Groundwater Investigation

Drilling and Installation of Monitor Wells	$6,000 per well or $150 per VLF (subcontractor cost - includes all Mob, Demob, Overhead, G&A, and Profit)
Technical Labor (includes Geologist, Technician, H&S Watch, QA, and Project Management)	50 workhours per well or 2 workhours per VLF
	50 workhours x $50/workhour (includes Overhead, G&A, and Fees)
	= $2,500 per well or $100 per VLF

   3. Water Sampling

Technical Labor (Technician)	0.70 workhours per sample
	0.70 workhours x $50/workhour (includes Overhead, G&A, and Fee)
	= $35 per sample

   4. Aquifer Testing (Slug Testing)

Technical Labor (includes Geologist and Technician)	10 workhours per well
	10 workhours x $50/workhour (includes Overhead, G&A, and Fee)
	= $500 per well

(*continues*)

Table 8.2 (*continued*)

5. Air Sampling

Hi-Vol/PM-10*	$90 per sample
Metals*	$90 per sample
Pesticides/PCBs*	$70 per sample
VOCs*	$70 per sample
Semi-VOCs*	$70 per sample
Meteorological*	$180 per sample

One-time Costs:

Equipment Preparation*	$45 per sampler
Transport*	$80 per sampler
Setup*	$30 per sampler
Teardown*	$30 per sampler

* Includes Labor and Equipment Rental Costs. Analysis Costs are shown in Part B. Costs above assume one hour per type of sample, including all sampling, flow checking, calibration, chain-of-custody, and documentation. Costs may be lower if several types of samples are being collected concurrently.

B. Analysis (Cost per Sample)

Parameter	Clear Water	Other Waters/Soils	Sludge/Other Solids
CLP Volatile Organics	$305	$350	$400
CLP VOC with 10 Compound Library Search	$355	$400	$450
CLP Acids + Base Neutral Combined	$550	$650	$750
CLKP Acids + Base Neutral Combined with 20 Compound Library Search	$600	$700	$800
CLP Organochlorine Pesticides & PCBs	$215	$240	$290
CLP Metals (Ag, Al, As, Ba, Be, Ca, Cd, Co, Cr, Cu, Fe, Hg, K, Mg, Mn, Na, Ni, Pb, Sb, Se, Tl, V, Zn)	$365	$450	$540
CLP Inorganics (Cyanides)	$150	$165	$185
Dioxin (2,3,7,8-TCDD)	$365	$440	$540
Total Dioxins	$700	$750	$850
Sulfate	$15	$25	$25
Gross Alpha/Beta	$40	$40	$40
Tritium	$55	N/A	N/A
Radiological Preparation Charge	$25	$25	$25

Table 8.2   (*continued*)

Parameter	Clear Water	Other Waters/Soils	Sludge/Other Solids
Pu-239	$130	$150	$150
Am-241	$130	$150	$150
Uranium Isotopes	$130	$150	$150
Sr-90	$100	$100	$100

Laboratory Analysis - Air Samples

Hi-Vol/PM10	$20 per sample
Metals	$400 per sample
VOC	$360 per sample
PUF-Pesticides/PCBs	$450 per sample
PUF-Semi-VOC	$450 per sample

NOTES:

1. Costs are in mid-1991 dollars.

2. Costs for Contract Laboratory Program (CLP) Analyses include dollars for CLP data report; costs for analyses with Certificate of Analysis report are approximately 10 percent lower.

3. Costs are for routine turnaround times of 10 to 15 working days, with the exception of Dioxin and Mixed Waste which are 20 working days. For rush or emergency service, add the following percentages:

Less than 48 hours	Cost + 200%
48 to 96 hours	Cost + 100%
5 to 10 days	Cost + 50%
Other	Cost + 25%

4. Costs shown are for one hypothetical sample. Volume discounts are available for a large number of samples. The volume discount is approximately 20 percent.

Sample Shipping (Overnight)

Coleman Cooler $75 per cooler

Each cooler will hold four to five surface water, groundwater, soil, and sediment samples.

Table 8.3 Estimated Purchase Costs for Sampling Equipment

Aquifer Testing Equipment	
Packer Test Setup	$3,000
Slug Test Setup	$100
Radiological Equipment	
Survey meter with probe	$530
Pancake probe	$200
Alpha scintillator	$1,250
Used as survey meter	$2,950
Direct reading dosimeters	$85/each
Charger	$91/each
TLD	$6.50/each
Air Monitoring Equipment	
HNu	$4,430
OVA	$6,600
LEL, 02 meter (MSA 260)	$1,653
Monitoring pumps	$975/each
Draeger pump	$207.65
Box of tubes/10@	$35
Miscellaneous	
Heat stress analyzer	$450
Fire extinguishers - ABC	$50
Decontamination setup: rubber tubs	$20/each
Brushes	$10/each
Detergent (i.e., TSP)	$10
Gatorade	$5/5 quarts
Field Equipment	
Self-contained breathing apparatus	$1,200/each
Fiddler	$5,000
Hand auger	$125
Bennet pump	$2,000
Stainless steel bailer	$150
Generator	$800
Sediment sampling	$100
Surface water sampler	$100

Bench studies differ from pilot studies in purpose, size, costs, and other criteria. Bench studies attempt to determine the effectiveness and feasibility of a particular technology over a wide range of conditions, whereas a pilot study focuses on a narrow range of selective technologies. Bench tests are normally performed in a laboratory environment; pilot studies are normally performed in the field. The activities associated with implementing either type of study are identical.

The first step in implementing a bench or pilot study is to develop a clear statement of work (SOW) that will define the objectives and level of detail for the study. The performance of the actual bench or pilot study will be based on the characteristics of the waste, the characteristics of the site, and the technologies being considered. The results of both the bench and pilot study are used in the feasibility study to determine the technical performance of the technology as well as to assist in estimating the cost of the full-scale process.

Although the costs of performing a bench or pilot study will vary based on site-specific conditions, the following rules of thumb may be used for estimating purposes:

- A bench study will range between 0.5% to 2% of the total capital cost of the alternative.
- Pilot studies will range between 2% to 5% of the capital cost of the alternative.

Data Analysis and RI Report

The final step of the RI process is the analysis of the data obtained and the development of the RI report. The draft RI report, produced at the end of the RI process, characterizes the site and summarizes the data collected and the conclusions drawn from the investigation activities. The report may be combined with the FS Report. After review, approval, and revisions, the draft report then becomes final.

The costs for this final phase of the RI cover labor to analyze the data and produce the report. The range of work hours for this phase was 3.2 to 120 work hours per hectare, with an average of 16.9 work hours per hectare. Assuming a rate of $50 per work hour results in a range of costs of $162 to $6,026 per hectare, with an average cost of $846 per hectare. These costs include data evaluation/analysis and report publication. Costs for travel and reproduction should be added as appropriate.

This section defines the normal activities associated with a FS. Because no data providing hours and costs for the individual tasks of the FS were readily available, total costs for four sites are provided at the end of this section.

Project Scoping

The first step of the FS process is to define the objectives of the remedial action and to broadly develop general responses to the problem. Modifications to the general response actions may be necessary as more data are collected or as the general purpose actions are more fully defined and developed.

Alternative Identification

The next step is to use the data from the RI to identify and develop a limited number of remedial action alternatives. Once the general response actions have been defined, feasible technologies will be identified to address the responses. Technologies that are clearly limited by both the site and waste characteristics should be eliminated. Those that have proven to be unreliable, are poor performers, or have not been fully demonstrated should also be eliminated. In general, the following alternatives are identified:

- Alternatives for treatment or disposal at an off-site facility.
- Alternatives that attain applicable and relevant federal public health or environmental standards.
- Alternatives that exceed applicable and relevant public health or environmental standards (as applicable).
- Alternatives that do not attain applicable and relevant public health and environmental standards, but will reduce the likelihood of present or future threat from the hazardous substance.
- A no-action alternative.

Alternative Screening

The next step in the FS process is to screen the technology alternatives that have been identified. The screening process occurs in two steps. The first step is to screen the alternatives based on environmental and public health criteria. An analysis will be performed to identify any adverse impacts on the environment and/or public health and welfare that may preclude the use of any alternatives.

The second step of the screening process is to screen alternatives based on order of magnitude cost. The purpose of this screening is to identify any alternatives that may be an order of magnitude higher than the other alternatives, but that do not provide additional environmental or public health benefits or increased reliability. The alternatives that receive the most favorable composite evaluation of public health, environmental, and cost factors are then documented and retained for further detailed technical evaluation.

Detailed Technical Evaluation

The alternatives selected in the screening process are now more fully defined to match them to the contaminated media. The alternatives are then evaluated against the five specific criteria outlined below:

- Short-term effectiveness
- Long-term effectiveness
- Reduction of toxicity, mobility, and volume
- Implementability
- Cost

Other items that are also evaluated are compliance with federal and state regulations, support agency and state acceptance, community acceptance, and the overall protection of human health and the environment.

The evaluation of the alternatives is then summarized to show the present worth of the total costs, health information, environmental effects, technical impacts, community impacts, and other factors.

FS Report

The final step of the FS process is the development of the FS Report. The draft FS Report will be subject to the support agency's review and comment, and possibly to other agencies' input before becoming final. The RI and FS reports are then used as the basis for selecting the appropriate remedial action technology and for documenting the evaluation process.

Other Activities

Data management, community relations, and health and safety planning activities also take place during the FS process. The activities are similar to those that take place during the RI process, with the exception of health and safety planning. Because no field activities occur during the FS, activities are those associated with safe working practices.

The range of costs shown in the RI for these activities may also be used for the FS.

Table 8.4   Summary of Total FS Costs

Site Location	Site Description	Hectares	Total FS Cost[a]	FS Cost per Hectare
Pennsylvania	Industrial Waste Lagoon	1.62	$82,189	$50,734
Nebraska	Processing plant, chemical soakage pit, underground storage tanks (USTs)	0.24	$26,225	$109,271
Colorado	Landfill containing waste from chemical processing facilities	4.05[b]	$68,185	$16,836
Missouri	TNT Production Facility	688.5	$251,760	$366

[a]   Actual costs for Pennsylvania and Nebraska; proposed costs for Colorado and Missouri.

[b]   Total Landfill Site.

Costs

Case studies were reviewed in order to develop total costs for the FS study. Four sites were highlighted and analyzed to determine costs for the FS process. A summary of the costs for these four sites is shown in Table 8.4.

8.2.2 Copper Mine Example

In the early 1970s, a high-grade copper ore deposit was discovered near Ladysmith, Wisconsin. Representative core samples reached 42% content, with a significant number of samples in the 15% to 20% range. Studies were performed to determine the most cost-effective techniques and recovery ratios for mining, crushing, milling, and concentrating of the ore and secondary precious metals. Plans were developed for the layout of facilities, utilities, and support services. Local opposition was encountered, mainly because of the need for the tailings pond required for the milling and concentrating operation. After numerous attempts at negotiation failed to resolve the issue, the plans were shelved.

In the early 1980s some of the local attitudes began changing. Wisconsin was foreseeing that agriculture and dairy interests alone were not sufficient to carry an ever-increasing cost of providing services to the citizenry of the area, and began looking at "clean" industries as a new revenue tax base source. It was then that the mining company altered its plans and proposed to mine the ore and primary-crush it to a size suitable for loading and shipping by rail out of state for milling, concentrating, and smelting at existing facilities. This new approach eliminated the need for a tailings pond on the site.

New plans were developed and submitted for preliminary approval. Upon first review, several low-lying areas were declared to be wetlands, and the configuration of the service facilities, roads, railspur, and incoming power routing was changed to accommodate wetlands regulations. Plans were resubmitted and a second determination was made that all water coming in contact with exposed ore and waste had to be treated prior to discharge in a nearby river. The operation of the sediment settling ponds and water treatment plant was designed to be a 24-hour per day, 7-day per week entity to treat in-pit water and surface water capacity, as calculated by normal inflows, precipitation patterns, and 10-year/50-year flood criteria.

The original plans designated that, upon completion of mining operations, the pit would be allowed to fill with water and that the adjacent land would be developed into a park site. In the final analysis, criterion that was deemed acceptable was that mined waste be returned to the pit in the reverse order of extraction so that silty clays, rock, and saphrolite would be restored to their original state insofar as possible. This approach would require a classification of wastes.

Waste would have to be segregated into separate piles, dependent upon sulfide content. Waste that would exceed 1% sulfide would require placement upon a lined pad with water containment dikes and a water collection system. The site was designed to be contoured to facilitate run-off of surface water into collection ponds. Water collected in the pit would be pumped to the sediment ponds, and subsequently to the water treatment plant. Areas under the rail loadout spot would require a paved surface, as would areas around the crushing plant and equipment maintenance shops. Because of the proximity to the river, extra precautions would be taken to curtail water inflow to the pit by installing a slurry wall. Also, because of the proximity to the townsite (schools, hospitals, convent), other restrictions would be imposed to allow only blasting, crushing, and ore loadout during weekday daylight hours. Entry into the plant site is via a state highway, further restricting receiving and shipments during off-peak hours.

To satisfy federal, state, and local regulations and ordinances, extensive background collection and laboratory analysis of plant life, animal and aquatic specimens, air quality, water quality (of the river and nearby residential water wells), and noise levels was required to establish a baseline. Public relations became a prominent concern, as reflected in the numerous meetings and forums with state committees, township councils, business leaders, and public interest groups. Depositions were recorded for public reviews and judicial hearings. All documentation was subject to disclosure.

An approximate cost of $5 million was expended over a two-and-one-half year period to fulfill obligations for permitting applications—all before one spade of dirt could be turned in pursuit of the mining venture. This time frame compared to an estimated life-of-mine of only 6 years' operation, plus 2 years of backfilling. The capital investment of site preparation, pre-stripping, mine development, water treatment, support services, and reclamation was estimated to be $30 million, including the cost of mining equipment. Hence, the venture spent 16.7% of the estimated capital investment and 31.3% of the expected mine life (plus reclamation) in attempting to apply for permits. The return on investment for such a venture must justify the owner bearing this magnitude of expense and time delay.

8.2.3 Gold Mine Example

Another example of up-front costs is related to a proposed gold mine located in the South Pacific where $25 million was spent on Feasibility Studies over a 5-year period. The studies concerned infrastructure development, mine development within a geothermal zone, waste disposal considerations, and political concerns.

8.2.4 Abandoned Mine Site Example

The once-abandoned mine shafts of the turn-of-the-century era are things of the past. Measures must now be taken to seal shafts and tunnels and to protect against subsidence; open pits must be dressed; sites declared to be hazardous must be cleaned up. The cost of this restoration is the responsibility of the owner/developer.

The scope of work at one such site is as follows:

- Flooding and sealing abandoned mine
- Demolition of buildings, structures, process circuits
- Decontamination of salvaged items
- Collapse and burial of nonsalvaged items
- Contouring and landscaping

* Neutralization and consolidation of tailings
* Isolation and warnings
* Systematic environmental monitoring.

The initial cost to develop the mine, to construct the buildings and the crushing/grinding/process circuits, to provide for storage of waste ore and tailings, and to develop the infrastructure to support the operation approximated $90 million. The mine produced a product for 3 years before it was shut down because of a downturn in market prices coupled with a reduction in demand. Several years have since passed, and now effort is under way to comply with state and federal agency regulations to limit contamination of the environment, particularly groundwater that would otherwise affect grazing, livestock, and ultimately human consumption. This effort may have an elapsed duration of 10 years at an additional computed cost of 15% to 17% of the original investment (approximately $13.5 to $15.3 million).

In addition to normal activities of demolition, the physical work includes such items as:

* Scrape site and haul roads
* Removing and/or containing contaminated soils, waste, and debris
* Creating diversion channels and lined ponds to capture water run-off for evaporation
* Farming of riprap from off-site sources for use as cap material
* Providing fenced enclosure with warning signs
* Providing pumpback wells and monitor wells for groundwater isolation and sampling
* Decontaminating of personnel and equipment upon exit of controlled site
* Documenting as-buried plot plans identifying coordinates, burial depth layer, and description of item.

An example of factors that influence the regulatory costs of reclamation to be considered in budget authorization include the following (a quantified cost breakdown is not authorized for publication). Mid-1986 pricing has been escalated to mid-1991 for consistency.

* Establish background criteria upon which to monitor compliance and effectiveness of reclamation techniques (including groundwater and soil monitoring characterization). This could be a 4-month exercise at a cost of some $1,050/day.
* Establish periodic procedure inspections at costs of $3,000 each.

• Establish routine alpha/gamma surveys at costs of $1,200 each.
• Establish bio-assay sampling: upwards of $58,000.
• Monitor site and construction personnel (specimen analysis) with cost ranges of:

Health physicist	$370/day
Health technician	$200/day
Analytical lab services	$24-37/each (uranium)
Analytical lab services	$78-225/each (radium)
TLD badges	$5/each

• Establish environmental monitoring; sampling of groundwater, air, and radon at costs of $46,200/year.
• Establish closure reviews: as much as $35,000.

8.2.5 Storage of Hazardous Materials

Another example of disposal costs relates to long-term storage of hazardous materials, such as flue dust that may be laden with arsenic and heavy metals. Involved in this operation are the costs of recovery, the costs of hauling in closed containers, and placement at approved disposal sites. Carriers of hazardous materials that travel on public roadways must be decontaminated after each trip. In most cases, this can be accomplished by steam cleaning in designated washdown areas where the rinse solution can be captured. More serious contamination may require sandblasting. For a point of reference, the approved depository was located some 80 miles from the site of origin. These sites must be constructed in such a manner as to eliminate the possibilities of groundwater contamination and the release of decomposition gases. Figure 8.5 is representative of an acceptable type of facility proposed.

The following sections provide costs for other environmental restoration activities. These costs were also derived from both actual experience and case studies. Again, the costs may be used to develop budget or check level estimates and are given in mid-1991 dollars. Many costs are not yet standardized in the environmental industry, due in part to the changeability of regulations and the uncertainty of government agency responsibilities. However, this book represents a major step in the publication of standardized costs. In any event, the scope of work for any environmental restoration activity must be defined carefully and consideration must be given to potential delays in the approval process for any project. The result of inspections and compliance monitoring may extend schedules and impact costs. A research of precedence in the RI/FS phase is time well spent.

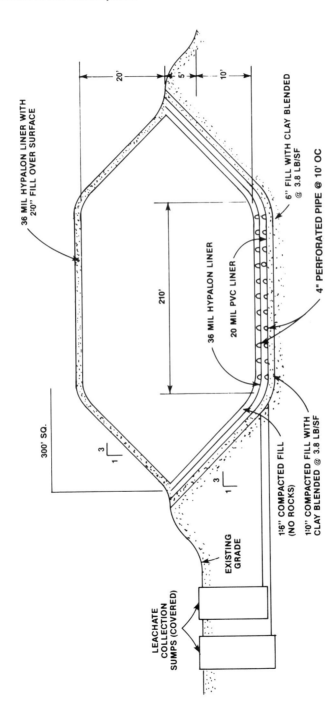

Figure 8.5 Typical cross section of flue dust depository.

8.2.6 Burial

Burial is one of the most widely used methods for disposing of contaminated solid materials. It is an acceptable disposal method for all types of solid materials including soils, mill tailings, rubble, and drummed waste. However, sludges must be solidified or stabilized before burial and liquids may not be buried. Burial is acceptable for the following contaminant types: chemicals, low-level radiation, and a mixture of both.

Contaminated materials can be buried in one of three types of facilities: on-site landfills, off-site landfills, and engineered caps. For large quantities, an on-site landfill could be constructed. Alternatively, if the contaminant will not migrate through the soil, an engineered cap may be sufficient. Engineered caps are typically used at sites contaminated only with low-level radioactivity. For smaller quantities, excavating the material and disposing it in an off-site landfill may be more cost effective. The U.S. Environmental Protection Agency (EPA) does not allow sludges or materials with free liquids to be buried without first reducing the liquid content. Typically, these materials are solidified by mixing them with fly ash, cement, kiln dust, etc. Slurry walls can also be installed to contain contaminated materials.

This section provides cost data for each of the disposal methods mentioned above.

On-Site Landfill

Description: The cost data presented here are based on the construction of an on-site landfill designed to hold 300,000 cubic yards of waste. The landfill would be located on a 15-acre rural site in the Gulf Coast region with gently rolling topography. The landfill is primarily aboveground; however, an average of 13 feet of cut-and-fill material will be excavated from the area prior to landfill construction.

The landfill complies with the RCRA requirements for a hazardous waste disposal facility. The bottom and side slopes are constructed of multiple layers of clay, flexible membrane liners (FMLs), sand, gravel, and porous geonet materials. (The geonet is a 250-mil thick synthetic layer formed in a net pattern that permits drainage in all directions. It is used for leak and leachate collection.) Whenever possible, on-site earth materials are used. The synthetic FMLs and geonets are made from high-density polyethylene (HDPE). The clay provides structural support and helps prevent liquids from migrating into or out of the landfill. The FMLs are the primary barriers against liquids migrating out of the landfill. In conjunction with the geonet, they also direct the flow of leachate toward the collection and removal system. The sand layers protect the FMLS against damage from heavy equipment during landfill construction and waste placement. Pea gravel is placed around the perforated pipes for the leachate collection and removal system (LCRS) to prevent the slots in the pipes from becoming plugged.

The landfill contains four waste cells. The LCRS collects any liquid that accumulates inside the landfill to prevent it from migrating through the liners and contaminating the groundwater. The liquid is transported to collection points. The LCRS consists of perforated pipes that follow the perimeter at the bottom of each waste cell. Any leachate is collected in the pipes and drains to one of four sumps at the bottom of manholes.

The LCRS is sized to collect 1,360 gallons per acre per day. The pipes are 6-inch diameter HDPE. The perforations are 0.10-inch slots. During operation and after closure of the landfill, any leachate that collects in the sumps is pumped into tank trucks and hauled to a treatment facility.

A leak detection, collection, and removal system (LDCRS) is located below the landfill to collect any liquids that escape. The LDCRS is constructed of the same materials as the LCRS. The collected liquids drain to one of four sumps, which are separate from those of the LCRS. Any liquids collected in the sumps are pumped into tank trucks and hauled to a treatment facility.

A groundwater monitoring system is also installed to verify the integrity of the landfill both during operation and after closure. The monitoring system provides the means to sample the groundwater upgradient and downgradient of the landfill. The samples are analyzed to determine if the groundwater is being contaminated by the landfill. The groundwater monitoring system includes ten new wells installed around the landfill's perimeter and four drains to permit sampling underneath the landfill. The wells are each about 50 feet deep.

The landfill also includes a system of pipes and vents for removing any gases that accumulate in the cells. Fifteen vents discharge gas collected by the 4-inch perforated polyethylene pipes just below the final cover of the landfill. Each vent includes a gas sampling port and an activated carbon canister.

The landfill also provides for control of the contaminated stormwater during waste placement. The waste is placed in the landfill in four cells. Temporary covers are placed over the waste to minimize the amount of stormwater that contacts the waste and therefore must be decontaminated. Stormwater that collects in active cells is pumped into a retention basin for treatment. Stormwater that collects in cells that do not yet contain waste is pumped out and discharged off-site without treatment.

Cost: The cost data presented here are based on proprietary in-house IT Corporation data. A detailed cost estimate was performed for the site described above. That cost estimate was then independently checked. The line items were grouped into six major categories and the cost of each category determined as a percentage of the total landfill cost. These percentages are shown in Table 8.5. Table 8.6 presents a list of items included in the detailed cost estimate for the landfill and their associated unit prices. The cost data are presented in mid-1991 dollars for the Beaumont, Texas area.

Engineered Cap

Description: An engineered cap, also called a *surface seal*, is placed over a contaminated area to prevent the contaminants from becoming airborne and/or to prevent surface water from infiltrating the disposal area and generating leachate. An engineered cap may be used instead of a full landfill only when the potential for the contaminants to migrate through the existing soil is negligible because of the type of contaminant and/or permeability of the existing soil. Engineered caps are frequently used at sites contaminated solely with low-level radioactivity; for example, to cover piles of uranium mill tailings. Low-level radioactive contaminants do not migrate through soil. They present a health risk when they become airborne. Covering the contaminated materials with an engineered cap seals the contaminants in place. An engineered cap can be as simple as a single layer of asphalt, clay, or concrete. It may also be a multiple-layered structure constructed of sand, clay, and synthetic liners, similar to the

Table 8.5   Cost Percentages for Major Categories of Landfill

Description	Percent of Total Cost
Site Work[a]	9.0
Landfill Structure	42.0
LCRS/LDCRS[b]	1.5
Groundwater Monitoring System	1.0
Gas Collection and Venting	0.5
Stormwater Control	2.0

[a]   Sitework includes stripping of the landfill site, temporary access roads, and a fence.

[b]   LCRS = Leachate collection and removal system
  LDCRS = Leak detection, collection, and removal systems

Table 8.6 Landfill Unit Prices[a]

Description	Unit of Measure	Quantity	Unit Price
Sitework			
Stripping	ACRE	15	$4,550
Common excavation	BCY[b]	42,000	$2.79
Rock excavation	BCY	62,600	$5.35
Common borrow excavation	BCY	32,000	$2.79
Fencing	LF	3,700	$19.04
Temporary traffic barriers	LF	2,700	$32.43
Access roads (10 feet wide)	LF	850	$49.27
Hydroseeding & fertilizing	SY	73,000	$1.40
Landfill Structure			
Embankment construction (rock fill)	CCY[c]	94,000	$8.08
Clay for bottom lining	CCY	72,000	$17.30
60-mil HDPE lining	SY	100,000	$6.05
Geonet (GN)	SY	79,000	$3.75
Geotextile (woven polyethylene, GT)	SY	78,000	$1.34
Sand (12" thick over GT & GN) and temporary cover	CCY	6,600	$18.93
On-site borrow placement	CCY	35,000	$2.73
Placement of on-site borrow for temporary cover	CCY	6,200	$2.73
Temporary cover	SY	37,000	$3.08
Clay for cover cap (24" thick)	CCY	25,000	$17.30
40-mil HDPE lining	SY	39,000	$4.03
Topsoil (6" over on-site borrow)	CCY	11,000	$4.61
Honeycomb erosion control	SY	65,000	$1.54
Geotextile for divider berms (woven polypropylene)	SY	2,600	$1.34
Sand bags for temporary cover	EA	13,000	$5.61
Headwall and riprap	EA	3	$1,475
LCRS/LDCRS			
LCRS sumps/pumps/miscellaneous	EA	4	$7,385
LDCRS sumps/pumps/miscellaneous	EA	4	$1,339
LCRS 6" perforated HDPE piping	LF	1,300	$25.31
LCRS 1¼" non-perforated HDPE piping	LF	400	$0.90
LCRS 2" non-perforated HDPE piping	LF	700	$5.81

(continues)

Table 8.6 (*continued*)

Description	Unit of Measure	Quantity	Unit Price
LCRS 6" non-perforated HDPE piping	LF	1,900	$22.80
LDCRS 6" perforated HDPE piping	LF	40	$25.31
LDCRS 6" non-perforated HDPE piping	LF	850	$22.80
LCRS pump station and tank	EA	1	$25,418
Groundwater Monitoring System			
Sand for groundwater (12")	CCY	5,900	$18.93
Groundwater valve vault	EA	4	$4,697
Groundwater 8" HDPE perforated pipe	LF	250	$44.56
Groundwater 8" HDPE non-perforated pipe	LF	1,100	$40.48
Cover strip drains	LF	9,200	$3.24
Cover 4" corrugated polyethylene drainage pipe	LF	1,500	$0.39
Cover 6" corrugated polyethylene drainage pipe	LF	4,000	$0.79
Ten Wells	VLF	500	$150
Gas Collection and Venting			
4" corrugated polyethylene gas collection pipe	LF	5,900	$0.39
Gas vent riser	EA	15	$2,071
Stormwater Control			
Clean stormwater 6" HDPE pipe	LF	450	$22.80
Clean stormwater 8" HDPE pipe	LF	180	$40.48
Clean stormwater 10" HDPE pipe	LF	190	$62.87
Clean stormwater pumps/miscellaneous	EA	3	$2,908
Contaminated stormwater 8" HDPE pipe	LF	2,000	$40.48
Contaminated stormwater 21" bituminous coated CMP	LF	2,200	$24.58
Contaminated stormwater pumps/miscellaneous	EA	2	$2,980
Waste Placement			
Placement of waste in landfill	CCY	230,000	$14.85
Placement of over-excavated material	CCY	70,000	$14.85

[a]   The data presented here were developed from proprietary in-house IT Corp. data.

[b]   Bank cubic yards

[c]   Compacted cubic yards

cap on a RCRA landfill. The cap is sloped to prevent stormwater from collecting on top.

Cost: Costs for many types of engineered caps, such as a single layer of asphalt, concrete or clay, are available from standard cost references, for example, the R.S. Means Company Building Construction Cost Data Manual.

Table 8.7 provides the unit costs for a three-layer surface seal system. From the bottom up, the layers include (1) an impermeable layer, (2) a middle drainage layer, and (3) a top cover that can be revegetated. (Revegetation costs are not included.) The impermeable layer consists of a minimum of 2 feet of compacted clay covered by an impermeable synthetic membrane liner at least 20 mils thick, and protected above by a 6-inch layer of clay or sand. The middle drainage layer consists of 6 inches of sand topped by 6 inches of gravel, and the top layer is 2 feet of earth.

Table 8.7 Three-Layer Surface Seal System Unit Costs

	Cost[a]	
	On-Site	Off-Site
Clay to fill in depressions in waste material		
• Borrowed clay	--	$8.37
• Backfill with a 300 hp bulldozer	$0.95	$0.43
• Compaction with a sheepsfoot roller	$1.14	$1.14
• Water wagon	$0.35	$0.35
TOTAL	$2.44/cy	$10.29/cy
Clay layer to cap		
• Borrowed clay placement of cap	—	$8.37
• Placement of cap	$2.73	$0.43
• Compaction	$1.71	$1.71
• Water wagon	$0.35	$0.35
• Fine finishing	$0.21	$0.21
TOTAL	$5.00/cy	$11.07/cy
Geotechnical testing	$353/day	$353/day
Synthetic liner (includes 10% for waste)		$0.08/sf
Drainage layer		
• Borrowed sand	$--	$5.49
• Borrowed gravel	--	$13.86
• Excavation of on-site sand or gravel	2.26	--
• Placing of sand or gravel	0.40	$0.40

[a] On-site and off-site refer to the source of the material used to construct the cap.

The design presented is consistent with the minimum requirements set forth in the draft "RCRA Guidance Document Landfill Design Liner Systems and Final Cover," July 1982.

The primary components of the surface seal module can include on-site excavation or purchase and delivery of sand, gravel, clay, and earth; placement of the aforementioned materials; geotechnical testing; and placement of synthetic membrane liner and geotextile fabric.

Off-Site Disposal

Description: Instead of constructing a landfill at the site, the waste can frequently be excavated and hauled to a commercial disposal facility. However, the off-site disposal facility must be constructed and permitted to accept the type of waste to be disposed. Prior to accepting wastes from a new customer, the disposal facility must analyze each waste stream to ensure that they are permitted to accept it. The analysis may be as simple as studying the material safety data sheet (MSDS) of off-specification products or as complex as sampling the waste and performing a full chemical analysis. The facility must also analyze new wastes from existing customers. Periodic analysis of ongoing waste streams may also be required.

In shipping the waste off-site, the risk of exposure to the waste increases because of the additional handling required. Also, construction of an on-site landfill is impractical and not cost efficient for small quantities of waste material. Hence, shipping to an off-site facility may be the most cost-effective approach.

Costs: The costs presented here are disposal facility costs only and do not include the excavation, loading, or transportation of the material to the facility. The disposal facility cost is comprised of the disposal fee and applicable surcharges. Table 8.8 lists the disposal fees and the most common surcharges. The unit "waste stream" in Table 8.8 refers to each time a commercial hazardous waste treatment or disposal facility receives a different waste, which must be tested to determine its hazardous constituents. Each waste can be considered a "waste stream" because of the different chemical and/or physical characteristics. As an example, the contaminated soil from two different sites would comprise two waste streams.

Stabilization/Solidification

Description: Stabilization and solidification are two similar processes with slightly different goals. In both processes the waste is mixed with reagents to generate a less hazardous substance by rendering it inert. Stabilization systems attempt to reduce the solubility or chemical reactivity of a

Table 8.8 Hazardous Waste Disposal Fees and Surcharges[a]

Disposal Fees	Unit	Non-PCB-Containing Waste	PCB-Containing Waste
Bulk Solids			
w/o stabilization	ton	$141-$188	$225
w/stabilization	ton	$225-$262	
Bulk Liquids/Sludges	gallon	$1.46	
Drummed Solids[b]			
w/o stabilization	55-gallon drum	$ 82-$ 89	$162
w/stabilization	55-gallon drum	$209	
Contaminated empty 55-gallon drum	each	$ 42	$ 42
Drummed Liquids[c,d,e]			
Organic, fuel[f]	55-gallon drum	$ 65	
Organic	55-gallon drum	$345	
Inorganic	55-gallon drum	$127	
Less than 100,000 ppm PCB	55-gallon drum		$262
Greater than 100,000 ppm PCB	55-gallon drum		$340
Drummed Sludge[g]			
Organic, fuel	55-gallon drum	$225	
Organic	55-gallon drum	$366	
Inorganic	55-gallon drum	$141	
Lab Packs	each	$262	

Surcharges	Unit	Non-PCB-Containing Waste	PCB-Containing Waste
Initial waste stream evaluation	waste stream	$314-$418	
Rebate on first shipment	waste stream	$209	
Drummed waste minimum	waste/shipment	$157-$209	
Bulk waste minimum			
Solids	shipment	$1,255	
Liquids	shipment	$2,301	
Truck washout	truck	$157	
Drain & flush transformer carcasses	CF		$10
Resolution of manifest	manifest	$21	$21

[a] Data are taken from vendor quotes of waste disposal for IT Corp. projects. Costs are presented in mid-1991 dollars.

[b] Drums containing solids and less than 90 percent full incur a $20 per drum surcharge.

[c] Drummed liquids are often over packed; i.e., the 55-gallon drum of waste is placed inside an 85-gallon drum and surrounded by an absorbent material.

[d] Over packed drums that are landfilled or solidified and landfilled incur a 155 percent surcharge.

[e] Frozen drums incur a $25 dollar per drum surcharge.

[f] Over packed drums used as fuel incur a $30 per drum surcharge.

[g] Closed top bung-type drums of sludge incur a $30 per drum surcharge.

189

waste by changing its chemical state or by physical entrapment. Solidification systems attempt to convert the waste into an easily handled solid with reduced hazards from volatilization, leaching, or spillage.

Both stabilization and solidification are applicable to the disposal of radioactive waste. In fact, many of the solidification process developments originated in low-level radioactive waste disposal. Some types of chemically hazardous waste disposal are compatible with stabilization and/or solidification. After a chemically contaminated waste is treated, an effort is often made to have the waste delisted by testing it for extraction procedure (EP) toxicity.

Most stabilization/solidification processes marketed are proprietary; often process changes are made to accommodate specific wastes. However, the more common process types include:

• Lime/fly ash pozzolan systems
• Pozzolan-portland cement systems
• Thermoplastic microencapsulation
• Surface encapsulation

In lime/fly ash pozzolan and pozzolan-portland cement processes, the waste is mixed with reagents to form a hard waste/concrete composite material. (Pozzolanic materials set to a solid when mixed with hydrated lime.) Thermoplastic microencapsulation involves mixing fine particulate waste with melted asphalt. After hardening, these materials can be buried with or without containers. Surface encapsulation (macroencapsulation) systems contain contaminants by placing an inert coating or jacket around a mass of cemented waste or by sealing them in polyethylene lined drums or containers. Table 8.9 shows the compatibility of each process type with several types of waste.

Cost: This section provides costs for five stabilization/solidification scenarios including:

• In-drum mixing
• In-situ mixing
• Mobile plant mixing of pumpable materials
• Mobile plant mixing of unpumpable materials
• Area mixing

In-drum mixing is best suited to highly toxic wastes present in relatively small quantities or when the waste is stored in drums with sufficient integrity to allow handling. However, complete mixing is difficult to achieve.

Table 8.9 Compatibility of Selected Waste Categories with Different Stabilization/Solidification Techniques[a]

Waste Component	Treatment Type			
	Cement-Based	Pozzolan-Based	Thermoplastic Microencapsulation	Surface Encapsulation
Organics				
Organic solvents and oils	May impede setting, may escape as vapor	May impede setting, may escape as vapor	Organics may vaporize on heating	Must first be absorbed on solid matrix
Solid organics (e.g., plastics, resins, tars)	Good—often increases durability	Good—often increases durability	Possible use as binding agent in this system	Compatible—many encapsulation materials are plastic
Inorganics				
Acid wastes	Cement will neutralize acids	Cement will neutralize acids	Can be neutralized before incorporation	Can be neutralized before incorporation
Oxidizers	Compatible	Compatible	May cause matrix breakdown, fire	May cause deterioration of encapsulation materials
Sulfates	May retard setting and cause spalling unless special cement is used	Compatible	May dehydrate and rehydrate causing splitting	Compatible
Halides	Easily leached from cement, may retard setting	May retard set, most are easily leached	May dehydrate and rehydrate	Compatible
Heavy metals	Compatible	Compatible	Compatible	Compatible
Radioactive materials	Compatible	Compatible	Compatible	Compatible

[a] Taken from the Handbook for Stabilization/Solidification of Hazardous Waste by U.S. Environmental Protection Agency; EPA/540/2-86-001.

In-situ mixing is best suited for closure of liquid or slurry holding ponds when large quantities of low-reactivity solid chemicals are added. Completel mixing at large sites is also difficult to achieve. Mobile mixing plants can be adapted to liquids, slurries, and solids. The waste material is pumped or excavated from the site and transported to the mixing plant. After the waste is mixed with the reagent, the material is shipped off-site or replaced in its original location for disposal. Mobile mixing plants provide the best quality control of all of the alternatives. Area mixing consists of spreading the waste and treatment reagents in alternating layers at the final disposal site and mixing them in place. Area mixing is not applicable for liquids and requires large load areas.

Table 8.10 presents a cost breakdown for each stabilization/solidification alternative based on mixing 68% waste, 30% portland cement, and

Table 8.10 Summary Comparison of Relative Costs for Stabilization/Solidification Alternatives[a,b]

Parameter	In-Drum	In-Situ	Plant Mixing		Area Mixing
			Pumpable	Unpumpable	
Metering and mixing efficiency	Good	Fair	Excellent	Excellent	Good
Processing days required	374	4	10	14	10
Cost/ton					
Reagent	$24.19	$24.19	$24.19	$24.19	$24.19
Labor and per diem	60.26	1.60	4.52	8.18	7.49
Equipment rental	43.82	1.63	4.63	8.90	4.80
Used drums @ $11/drum	56.85	--	--	--	--
Mobilization-demobilization	18.50	1.86	1.68	2.67	1.41
Cost of treatment process	203.64	29.29	35.03	43.94	37.90
Profit and overhead (30%)	61.09	8.79	10.51	12.62	11.37
Total cost/ton	$468.35	$67.36	$80.56	$100.50	$87.16

[a] Taken from the "Handbook of Stabilization/Solidification of Hazardous Wastes" by the U.S. Environmental Protection Agency; EPA/540/2-86-001.

[b] In all cases, 500,000 gal (2,850 tons) of waste was treated with 30% portland cement and 2% sodium silicate with on-site disposal; costs include only those operations necessary for treatment. All costs are per ton of waste treated, mid-1991 dollars.

2% sodium silicate. Table 8.11 presents the cost per ton based on four other waste and reagent mixtures. Table 8.12 presents the purchase price of nine reagents often used for stabilization/solidification. All of the costs are in mid-1991 dollars.

Slurry Walls

Description: Slurry walls are sometimes used to prevent lateral migration of contaminants through the soil. If slurry walls are installed in the bedrock below a contaminated area, the contaminated area is, in effect, sealed in place. Slurry walls are typically made of soil/bentonite or cement/bentonite mixtures. A trench is dug and the mixture is poured into it without form work. Steel reinforcing is sometimes used.

Cost: Table 8.13 presents ranges of typical slurry wall costs as a function of the type of soil and depth of the trench.

Table 8.11   Comparison of Treatment Costs with Different Reagents[a,b]

Reagent Type, Amount, & Cost	In-Drum	In-Situ	Plant Mixing		Area Mixing
			Pumpable	Unpumpable	
1. 80% fly ash (Type F) @ $35.56/ton, 20% lime @ $58.58/ton. Total reagent cost/ton of waste = $35.56					
Total Cost/ton	$279.73	$58.87	$66.25	$77.82	$69.98
2. 30% Portland cement @ $64.85/ton, 2% sodium silicate @ $235.35/ton. Total reagent cost/ton of waste = $24.19					
Total cost/ton	$264.72	$38.07	$45.66	$57.12	$49.27
3. 50% fly ash (Type C) @ $24.05/ton. Total reagent cost/ton of waste = $11.51					
Total cost/ton	$247.68	$21.99	$29.45	$41.00	$33.09
4. Free reagent (including delivery) Total reagent cost/ton of waste = $0					
Total cost/ton	$234.31	$6.64	$14.10	$25.67	$17.81

[a]   Costs were recalculated for different reagent cost, but for the same equipment, project duration, and mobilization costs. All reagent proportions are in weight of reagent per weight of waste. Costs are in mid-1991 dollars.

[b]   Taken from the "Handbook for Stabilization/Solidification of Hazardous Waste" by the U.S. Environmental Protection Agency; EPA/540/2-86-001.

Table 8.12   Typical Costs of Chemicals Used for Stabilization/Solidification (Mid-1991)[a,b]

Chemical	Units	Cost range
Portland cement	$/ton[b] (bulk)	$47.00-78.00
Portland cement	$/ton (bag)	$84.00-99.00
Quick lime (Ca0)	$/ton (bulk)	$52.00-63.00
Hydrated lime (Ca(OH)$_2$)	$/ton (bulk)	$52.00-63.00
Hydrated lime (Ca(OH)$_2$)	$/ton (bag)	$68.00-89.00
Cement kiln dust	$/ton	$6.00-29.00
Waste quick lime	$/ton	$5.00-14.00
Fly ash	$/ton	$0.00-47.00
Gypsum	$/ton	$0.00-42.00
Sodium silicate	$/pound	$0.06-0.24
Concrete admixtures	$/gallon	$1.78-11.00

[a]   Taken from the "Handbook for Stabilization/Solidification of Hazardous Waste" by the U.S. Environmental Protection Agency; EPA/540/2-86-001.

[b]   Customary units are used because price quotations are made in these units. All prices F.O.B. at point of manufacture.

Table 8.13   Relation of Slurry Cut-Off Wall Costs per Square Foot as a Function of Medium and Depth[a]

Parent Material	Slurry trench prices, Soil bentonite backfill (Dollars/square foot) [d,e,f]			Unreinforced slurry walls prices, Cement bentonite backfill (Dollars/square foot) [d,e,f]		
	Depth ≤ 30 feet	Depth 30-75 feet	Depth 75-120 feet	Depth ≤ 60 feet	Depth 60-150 feet	Depth > 150 feet
Soft to medium soil N[b] ≤ 40	3-6	6-13	13-16	23-31	31-47	47-115
Hard soil N = 40-200	6-10	7-16	16-31	39-47	47-63	63-146
Occasional boulders	6-13	7-13	13-39	31-47	47-63	63-131
Soft to medium rock N ≥ 200 sandstone, shale	9-19	16-31	31-78	78-94	94-131	131-272
Boulder strata	23-39	23-39	78-126	47-63	63-146	146-324
Hard rock granite, gneiss, schist[c]	--	--	--	146-220	220-272	272-366

[a]   Costs are in Mid-1991 dollars (taken from EPA-540/2-84-001, page 7-24).

[b]   N is the standard penetration value in number of blows of the hammer per foot of penetration (ASTM D1586-67).

[c]   Normal penetration only.

[d]   For standard reinforcement add $12.55 per square foot.

[e]   For construction in an urban environment add 25 to 50 percent.

[f]   Square foot refers to a vertical square foot, trench length times trench depth. (Nominal width of trench ranges from 1 to 3 feet).

8.2.7   Building/Equipment Decontamination

Introduction

There are many decontamination methods that may be applied to buildings, equipment, and structures on inactive sites. Decontamination of these items is important to limit and prevent the spread of contamination off-site, as well as to reduce exposure levels to future users of the buildings or equipment. A successful decontamination application can also reduce the need for dismantling and disposing of contaminated structures. The value of the reconditioned buildings and equipment can also be increased or salvaged. The objective of the decontamination program is to return the buildings and equipment to a productive usable status or to minimize the volume of hazardous waste to be disposed of.

This section will describe the various methods of decontaminating buildings, structures, and equipment and will provide costs for each method. The source of information and cost data for the decontamination methods is the document, "Guide for Decontaminating Buildings, Structures, and Equipment at Superfund Sites," PEI Associates, Inc., and Battelle Columbus Laboratories, Hazardous Waste Engineering Research Laboratory, EPA/600/2-85/028, March 1985. The costs for the decontamination methods are presented on a cubic foot basis for a two-story model building which is 60 feet long, 30 feet wide, and 25 feet high. Each of the two stories has 1,800 square feet of floor area, 1,800 square feet of ceiling area, and the building's total wall area is 4,500 square feet. The floor and wall slabs are constructed of concrete; the walls and foundation are 1 foot thick and the floor of the secondary story is 4 inches thick. The building contains 5 tons of steel, including one boiler, 60 feet of piping, and stairs. The total volume of the building is 53,568 cubic feet. Waste disposal costs are assumed to be included in the operating costs, unless the waste generated is hazardous and must be incinerated or disposed of in a secure landfill.

A summary of the cost per cubic foot for each decontamination method is shown in Table 8.14. The costs are in mid-1991 dollars.

Table 8.14 Building/Equipment Decontamination Summary

Decontamination Method	Cost per cf[a]
Demolition	$1.43
Dismantling	$0.11
Dusting, wiping, vacuuming	$0.08
Gritblasting	$1.17
Hydroblasting	$2.95
Painting/coating	$0.07
Scarification	$1.29
Steam cleaning	$0.49
Acid etching	$0.22
Bleaching	$0.35
Flaming	$0.07
Drilling and spalling	$2.37

[a] Mid-1991 dollars; cost per cf for model building.

Decontamination Methods

Demolition: Demolition is the total destruction of a building, structure, or piece of equipment. Demolition is potentially applicable to all contaminants, and the technology is well developed. However, large quantities of contaminated debris must be disposed of and airborne contamination may occur during the demolition process. Other drawbacks to this method include the considerable building preparation activities that must take place (all surfaces must be washed down to minimize dust, and contaminant residues may have to be neutralized or stabilized to prevent explosions or emissions). Some structures inside the building may have to be dismantled prior to the demolition.

Equipment required for demolition includes explosives or other demolition equipment, clean-up equipment, water hoses, and personal protective gear. Personnel hazards include high noise and dust levels, as well as the explosives.

The cost for the individual tasks associated with demolition are shown in Table 8.15. Cleanup will comprise the largest portion of the "demolition" activity. The costs associated with the disposal of the debris comprise 81% of the total cost. Incineration costs increase as the noncombustible content of the debris increases. No costs are included for the construction of new facilities.

Dismantling: Dismantling is the physical removal of selected structures or equipment from inside a building and is potentially applicable to

Table 8.15   Demolition Costs[a]

Work component	Quantity	Cost[b]	Unit cost[b]	Percent of total
Demolition & cleanup[c]	53,568 cf	$13,033	$0.24/cf	17
Landfilling of debris	6,900 cf	$62,363	$9.04/cf	81
Personal protective equipment	NA	$1,161	—	2
Total	53,568 cf	$76,557	$1.43/cf[d]	100

[a]   Taken from "Guide for Decontaminating Buildings, Structures, and Equipment at Superfund Sites" prepared by PEI Associates, Inc., and Battelle Columbus Laboratories for U.S. Environmental Protection Agency, 1984.

[b]   Mid-1991 dollars.

[c]   Includes labor, equipment, materials, overhead, and profit.

[d]   Based on building volume before demolition.

all types of contamination. Dismantling may serve as the entire decontamination process or it may be the initial step of demolition. Dismantling is less costly than demolition, but still requires the disposal of large quantities of contaminated debris. The technique has been widely used and will result in complete decontamination of the structures and equipment removed.

The components to be dismantled must be identified, followed by the isolation of the work area. Tools and equipment required for the process include saws, wrecking bars, water sprayers, air compressor, safety equipment, and disposal supplies. Safety hazards are the same as those associated with the use of machinery and tools.

The costs for individual dismantling activities are shown in Table 8.16. Personnel, equipment, and materials make up the majority of the dismantling costs. Cleanup after the dismantling accounts for 36% of the total cost.

Dusting/Vacuuming/Wiping: This technique refers to the use of common cleaning methods to physically remove hazardous dust and particulates from building and equipment surfaces. Dusting/vacuuming/wiping is applicable to all types of particulate contamination and use of this method allows containment and easy disposal of the wastes. How-

Table 8.16 Dismantling Costs[a]

Work Component	Quantity	Cost[b]	Unit Cost[b]	Percent of Total
Dismantle boiler[c]	1 each	$1,726	$1,726/each	29
Dismantle piping[c]	60 lf	$160	$ 2.67/lf	3
Dismantle stairs[c]	20 risers	$345	$17.25/riser	6
Cleanup building	53,568 cf	$2,090	$0.04/cf	36
Landfill - debris	5 tons	$345	$69/ton	6
Personal protective equipment	—	$1,161	—	20
Total	53,568 cf	$5,827	$0.11/cf[d]	100

[a] Taken from "Guide for Decontaminating Buildings, Structures, and Equipment at Superfund Sites" prepared by PEI Associates, Inc., and Battelle Columbus Laboratories for U.S. Environmental Protection Agency, 1984.

[b] Mid-1991 dollars.

[c] Includes labor, equipment, materials, overhead, and profit.

[d] Based on building volume before demolition.

ever, the fugitive dust created by vacuuming may cause spreading of the contamination.

The equipment required for this method is: commercial vacuums with a high-efficiency particulate air (HEPA) filter, damp cloths or wipes soaked with solvent, and brushes and brooms for coarse debris. Toxic dust represents the primary safety hazard and workers should wear appropriate protective clothing and respirators.

The costs for the individual components of the method are shown in Table 8.17. The majority of the cost, 67%, is comprised of the labor to vacuum, dust, and wipe the building and equipment surfaces.

Gritblasting: Gritblasting is a surface removal decontamination technique that uses an abrasive material such as sand, steel pellets, alumina, or glass beads. Gritblasting generates large amounts of dust and debris and it is only effective as a surface treatment. Obstructions may require removal prior to gritblasting; the debris is collected and placed in appropriate containers for treatment or disposal. The building must then be vacuumed or washed with water.

The equipment required for this method includes a blast-gun, pressure lines, abrasive, and an air compressor. Workers should wear protective clothing and respirators to minimize the danger of dust inhalation.

The costs of the individual tasks for gritblasting are shown in Table 8.18. This method is labor intensive and disposal costs can be high if the debris is considered hazardous.

Table 8.17   Dusting/Vacuuming/Wiping/Costs[a]

Work Component	Quantity	Cost[b]	Unit Cost[b]	Percent of Total
Vacuuming, dusting, wiping[c]	3,600 sf	$2,772	$0.77/sf	67
Hand held vacuum	—	$188	—	5
Personal protective equipment	—	$1,161	—	28
Total	53,568 cf	$4,121	$0.08/cf[d]	100

[a]   Taken from "Guide for Decontaminating Buildings, Structures, and Equipment at Superfund Sites" prepared by PEI Associates, Inc., and Battelle Columbus Laboratories for U. S. Environmental Protection Agency, 1984.

[b]   Mid-1991 dollars.

[c]   Includes labor, equipment, materials, overhead, and profit.

[d]   Based on building volume before demolition.

Table 8.18 Gritblasting Costs[a]

Work Component	Quantity	Cost[b]	Unit Cost[b]	Percent of Total
Gritblast - walls & floors[c]	8,100 sf	$19,429	$2.40/sf	31
Landfilling of debris	3,008 cf	$41,830	13.91/cf	67
Personal protective equipment	—	$1,161	—	2
Total	53,568 cf	$62,420	$1.17/cf[d]	100

[a] Taken from "Guide for Decontaminating Buildings, Structures, and Equipment at Superfund Sites" prepared by PEI Associates, Inc., and Battelle Columbus Laboratories for U. S. Environmental Protection Agency, 1984.

[b] Mid-1991 dollars.

[c] Includes labor, equipment, materials, overhead, and profit.

[d] Based on building volume before demolition.

Hydroblasting: Hydroblasting uses a high-pressure water jet that removes contaminated debris from surfaces. This technique is applicable to explosives, heavy metals, and radioactive contaminants. Hydroblasting may not effectively remove contaminants that have penetrated beneath the surface and large amounts of contaminated liquids must be collected and treated.

Equipment used for hydroblasting includes off-the-shelf high-pressure pump hoses, nozzles, water collection sumps, water storage tanks, and water pumps. There are no process hazards associated with the method, but workers should wear protective clothing, glasses, gloves, and earplugs.

The costs for the individual components of hydroblasting are shown in Table 8.19. The costs are based on the assumption that the required equipment (hydroblaster, sump pump, storage tanks) would be purchased. These purchase costs represent 49% of the total cost. The labor to hydroblast the building represents 36% of the total cost.

Painting/Coating: This general decontamination method covers three specific techniques: (1) lead-based paint removal, (2) fixative/stabilizer coatings, and (3) strippable coatings. Paint removal uses commercially available paint removers or physical means such as scraping, scrubbing, or water washing. Surfaces are then repainted. The method is effective for all painted surfaces, but is very labor intensive. Various materials, such as molten and solid epoxy paint films and polyester resins, can be

Table 8.19   Hydroblasting Costs[a]

Work Component	Quantity	Cost[b]	Unit Cost[b]	Percent of Total
Setup tanks/pumps	—	$5,544	—	3
Setup equipment	—	$2,092	—	1
Hydroblasting - building	8,100 sf	$56,327	$6.95/sf	36
Hydroblasting - equipment	1 ea	$5,544	$5,544/ea	3
Cleanup	—	$2,772	—	2
Landfilling of debris	507 cf	$7,050	$13.91/cf	4
Personal protective equipment	—	$1,161	—	1
Hydroblaster[c]	—	$31,197	—	20
Pipe and tank cleaning accessories[c]	—	$7,887	—	5
Sump pump[c]	—	$1,004	—	1
Storage tanks[c]	—	$37,646	—	24
Total	53,568 cf	$158,224	$2.95/cf[d]	100

[a]   Taken from "Guide for Decontaminating Buildings, Structures, and Equipment at Superfund Sites" prepared by PEI Associates, Inc., and Battelle Columbus Laboratories for U.S. Environmental Protection Agency, 1984.

[b]   Mid-1991 dollars.

[c]   Includes labor, equipment, materials, overhead, and profit.

[d]   Based on building volume before demolition.

used as a coating to fix or stabilize contaminants in place. The technique is potentially applicable to all building materials, but documented use has been found only for PCBs, explosives, and radioactive contaminants. A polymer may also be used to bind with the contaminant, thereby making it easier to handle and for disposal.

Equipment required for these methods includes hand-scraping tools, brushes, paint removers, mixing equipment, storage tanks, and spraying equipment. Safety hazards are the possibility of exposure to airborne contaminants, solvent flammability and toxicity, and skin contact.

Table 8.20 shows the costs of the individual activities associated with painting/coating. No costs are included for paint removal and it was assumed that the walls would be the only coated area. The costs represent application of three coats (one coat primer, two coats semigloss) with a sprayer. It was also assumed that the sprayer equipment would be rented. The labor to paint/coat the walls accounts for 48% of the total cost.

Table 8.20   Painting/Coating Costs[a]

Work Component	Quantity	Cost[b]	Unit Cost[b]	Percent of Total
Setup/teardown equipment	—	$277	—	7
Paint/coat walls	47,075 sf	$1,883	$0.04/sf	48
Sprayer rental	4,500 cf	$146	—	4
Paint (primer)[c]	—	$146	—	4
Paint (semigloss)[c]	—	$293	—	7
Personal protective equipment	—	$1,161	—	30
Total	53,568 cf	$3,906	$0.07/cf[d]	100

[a] Taken from "Guide for Decontaminating Buildings, Structures, and Equipment at Superfund Sites" prepared by PEI Associates, Inc., and Battelle Columbus Laboratories for U.S. Environmental Protection Agency, 1984.

[b] Mid-1991 dollars.

[c] Includes labor, equipment, materials, overhead, and profit.

[d] Based on building volume before demolition.

Scarification: This method uses a tool consisting of pneumatically operated piston heads that strike the surface causing the concrete (or similar material) to chip off. It is capable of removing up to 1 inch of surface layer and is applicable to all contaminants except highly toxic residues and highly sensitive explosives. It is applicable only to concrete (not concrete block). Large amounts of contaminated water and concrete are generated and scarification leaves the area with a rough appearance, which must be resurfaced.

Equipment required for scarification are: a scarification tool, a portable generator, air compressors, a debris-collecting/packing system, and the tungsten-carbide bits for the tool. Personnel hazards include high noise, contaminant-laden dust, and injury from flying chips and vibration.

The costs for the components of scarification are shown in Table 8.21. The removal rate of a seven-piston floor scarifier is 315 square feet per hour, while the rate for a three-piston wall scarifier is 72 to 108 square feet per hour. The removal rates represent a removed depth of 1 inch. Because scarification leaves a rough and uneven surface, costs for wall and floor patching are required. The patching of floors and walls represents 63% of the total cost.

Table 8.21 Scarification Costs[a]

Work Component	Quantity	Cost[b]	Unit Cost[b]	Percent of Total
Scarification/cleanup walls[c]	4,500 sf	$4,796	$1.07/sf	7
Scarification/cleanup floors[c]	3,600 sf	$873	$0.24/sf	1
Patching - walls[c]	4,500 sf	$22,683	$5.04/sf	33
Patching - floors[c]	3,600 sf	$20,606	$5.72/sf	30
Landfilling - debris	1,350 sf	$18,770	$13.90/sf	27
Personal protective equipment	--	$1,161	—	2
Total	53,568 cf	$68,889	$1.29/cf[d]	100

[a] Taken from "Guide for Decontaminating Buildings, Structures, and Equipment at Superfund Sites" prepared by PEI Associates, Inc., and Battelle Columbus Laboratories for U.S. Environmental Protection Agency, 1984.

[b] Mid-1991 dollars.

[c] Includes labor, equipment, materials, overhead, and profit.

[d] Based on building volume before demolition.

Steam Cleaning: Steam cleaning uses hand-held wands or automated systems to extract contaminants physically from building and equipment surfaces. It is used mainly to remove contaminated particulates and explosive residues and is applicable to a variety of structural materials. It is known to be effective only for surface decontamination and is labor intensive.

Steam cleaning requires steam generators, spray systems, collection sumps, and waste treatment systems for the condensate. Safety hazards associated with this decontamination method are steam burns and toxicity of the steam and solvent mixtures.

The costs for steam cleaning are shown in Table 8.22. The rental cost is for a 200-gallon per hour steam cleaner; it is assumed that the sump pump and storage tank would be purchased. Disposal of the hazardous liquid accounts for 36% of the total cost; purchase of the storage tank represents 31% of the total cost.

Acid Etching: Acid etching involves the application of acid to the contaminated surface, which promotes corrosion and removal of the surface layer. The debris is then neutralized and disposed of. Acid etching is only a surface treatment and is applicable primarily to mild steel and wood surfaces and many contaminants. The technique is hazardous and requires

Table 8.22   Steam Cleaning Costs[a]

Work Component	Quantity	Cost[b]	Unit Cost[b]	Percent of Total
Setup/teardown equipment	—	$785	—	3
Steam clean - walls and floors[c]	8,100 sf	$5,622	$0.69/sf	21
Steam cleaner - rental	—	$513	—	2
Disposal - hazardous liquid	—	$9,383	—	36
Personal protective equipment	—	$1,161	—	4
Sump pump[c]	—	$811	—	3
Storage tank[c]	—	$8,112	—	31
Total	53,568 cf	$26,387	$0.49/cf[d]	100

[a]   Taken from "Guide for Decontaminating Buildings, Structures, and Equipment at Superfund Sites" prepared by PEI Associates, Inc., and Battelle Columbus Laboratories for U.S. Environmental Protection Agency, 1984.

[b]   Mid-1991 dollars.

[c]   Includes labor, equipment, materials, overhead, and profit.

[d]   Based on building volume before demolition.

special equipment, such as spraying equipment, a pump, an acid source, an acid neutralizer, and an optional stream source.

If the equipment used is not corrosion resistant, it will require considerable maintenance and periodic replacement. Personnel hazards are acid skin burns and the inhalation of toxic fumes.

The costs for acid etching are shown in Table 8.23. The costs assume that a commercial technique of acid etching/washing would be used, which eliminates the need for a neutralizing wash after the acid wash. A simple water rinse is included. The pump is used to transfer the liquid from the sump to drums for disposal and is assumed to be purchased. The acid wash/rinse of the walls and floors represents 44% of the total cost; disposal of the hazardous liquid accounts for 43% of the total cost.

Bleaching: Bleaching has been used against chemical agents and liquid pesticide spills. It is most effective on metal surfaces, but can be used on wood and concrete surfaces as well. Bleaching entails the application of a bleach formulation to the contaminated surface, where it is

Table 8.23 Acid Etching Costs[a]

Work Component	Quantity	Cost[b]	Unit Cost[b]	Percent of Total
Acid wash & rinse - walls & floors [c]	8,100 sf	$5,162	$0.64/sf	44
Disposal - hazardous liquid	361 cf	$5,021	$13.91/cf	43
Personal protective equipment	—	$1,161	—	10
Sump pump	—	$345	—	3
Total	53,568 cf	$11,689	$0.22/cf[d]	100

[a] Taken from "Guide for Decontaminating Buildings, Structures, and Equipment at Superfund Sites" prepared by PEI Associates, Inc., and Battelle Columbus Laboratories for U.S. Environmental Protection Agency, 1984.

[b] Mid-1991 dollars.

[c] Includes labor, equipment, materials, overhead, and profit.

[d] Based on building volume before demolition.

allowed to react with the contaminants and is then removed. It is primarily used in conjunction with other decontamination methods such as absorbents and/or water washing. The bleach formulation is generally applied as a slurry, which can lead to clogging of the application equipment.

The required equipment for bleaching includes hoses, scrubbers, containers, a waste recovery system, and safety equipment. Safety hazards result from possible exposure to chemical agents, as well as exposure to the contaminants.

The costs for the bleaching technique are shown in Table 8.24. The labor to apply the bleach formulation, scrub the surface, and rinse it comprises the majority of the costs (60%). Disposal of the rinse liquid, which is hazardous, is the next highest component at 27%.

Flaming: Flaming is the application of controlled, high-temperature flames to noncombustible surfaces in order to degrade the contaminants thermally. It can be used for all explosives and some low-level radioactive contaminants, and can be applied to both painted and unpainted concrete, cement, brick, and metals. It is primarily a surface decontamination technique, may detonate combustible residues, and can involve high fuel costs.

The use of flaming requires a torch (either hand-held or remotely operated), a fuel source, hoses, regulators, fire extinguishers, and tools to remove

Table 8.24 Bleaching Costs[a]

Work Component	Quantity	Cost[b]	Unit Cost[b]	Percent of Total
Bleach floors & walls[c]	8,100 sf	$11,271	$1.39/sf	60
Bleach solution	900 gal	$941	$1.05/gal	5
Disposal of hazardous liquid	361 cf	$5,021	$13.91/cf	27
Personal protective equipment	—	$1,161	—	6
Sump pump	1 ea	$345	$345/ea	2
Total	53,568 cf	$18,739	$0.35/cf[d]	100

[a] Taken from "Guide for Decontaminating Buildings, Structures, and Equipment at Superfund Sites" prepared by PEI Associates, Inc., and Battelle Columbus Laboratories for U.S. Environmental Protection Agency, 1984.

[b] Mid-1991 dollars.

[c] Includes labor, equipment, materials, overhead, and profit.

[d] Based on building volume before demolition.

Table 8.25 Flaming Costs[a]

Work Component	Quantity	Cost[b]	Unit Cost[b]	Percent of Total
Setup/teardown equipment	—	$209	—	6
Flame - walls & floors[c]	8,100 sf	$1,464	$0.18/sf	40
Cleanup (rinse)	—	$706	—	19
Flamer rental	—	$105	—	3
Personal protective equipment	—	$1,161	—	32
Total	53,568 cf	$3,645	$0.07/cf[d]	100

[a] Taken from "Guide for Decontaminating Buildings, Structures, and Equipment at Superfund Sites" prepared by PEI Associates, Inc., and Battelle Columbus Laboratories for U.S. Environmental Protection Agency, 1984.

[b] Mid-1991 dollars.

[c] Includes labor, equipment, materials, overhead, and profit.

[d] Based on building volume before demolition.

obstructions and combustible materials. Safety hazards are the production of gaseous pollutants that may require scrubbing to prevent release to the atmosphere, toxic vapors, and possible detonation of combustible materials.

The costs for the activities associated with flaming are shown in Table 8.25. The labor to flame the building surfaces and then rinse them comprises 60% of the total cost. The cost also assumes that the flamer or torch is rented.

Drilling and Spalling: Drilling and spalling entails drilling holes into the contaminated surface, inserting a spalling tool bit into the holes that then hydraulically spreads to spall or break off the contaminated concrete. It is applicable to concrete only (not concrete block). It can achieve deeper penetration than other surface removal methods but will result in a surface that is very coarse and that may require resurfacing.

A drilling/spalling rig, a scaffolding/hydraulic positioning system, and cleaning equipment are required to use the method. Personnel hazards are represented by high dust and noise levels, high-pressure air lines, and by flying debris.

Table 8.26   Drilling and Spalling Costs[a]

Work Component	Quantity	Cost[b]	Unit Cost[b]	Percent of Total
Setup/teardown	—	$836	—	1
Drill/spall - walls	6,300 sf	$32,656	$5.18/sf	26
Front-end loader	12 days	$6,228	$519/day	5
Demolish - second floor	1,800 sf	$4,645	$2.58/sf	4
Reconstruct - second floor	1,800 sf	$2,707	$1.50/sf	2
Patch first floor[c]	1,800 sf	$10,293	$5.72/sf	8
Patch walls[c]	4,500 sf	$22,661	$5.04/sf	17
Landfill debris	1,650 cf	$22,917	$13.89/cf	18
Personal protective equipment	—	$1,160	—	1
Drilling/spalling rig	1 ea	$23,147	$23,147/ea	18
Total	53,568 cf	$127,250	$2.38/cf[d]	100

[a]   Taken from "Guide for Decontaminating Buildings, Structures, and Equipment at Superfund Sites" prepared by PEI Associates, Inc., and Battelle Columbus Laboratories for U.S. Environmental Protection Agency, 1984.

[b]   Mid-1991 dollars.

[c]   Includes labor, equipment, materials, overhead, and profit.

[d]   Based on building volume before demolition.

The costs for drilling/spalling are shown in Table 8.26. The second floor is 4 inches thick, which is too thin to spall. This floor is demolished. The labor to drill and spall represents 26% of the total cost. The cost for patching the floors and walls accounts for 26% of the cost. It is also assumed that the drilling/spalling rig is purchased.

Other Decontamination Techniques: Solvent washing entails the circulation of an organic solvent across the building surface to solubilize the contaminants. It has potential applicability to a wide range of contaminants, depending on the solvent used. The method has not been widely used for buildings and requires further development in application, recovery, collection, and efficiency.

Vapor-phase solvent extraction is the heating to vaporization of an organic solvent that is then allowed to permeate throughout the building. The vapor permeates into porous surfaces, condenses, and solubilizes the contaminants. The contaminated solvent is then collected in a sump and treated. It has not yet been applied to building decontamination and the primary problem with it is achieving an outward flux of the contaminated solvent from the porous building surfaces.

K-20 Sealant is a commercially developed product that penetrates the porous surface and immobilizes contaminants in place. It is manufactured by Lopat Enterprises, Inc. Its effectiveness as a permanent barrier and hence as a potential application had not been verified as of March 1985.

Photochemical and microbial degradation are the use of ultraviolet (UV) and microbes, respectively, to degrade the contaminant. Neither has been proven effective as a method of building decontamination.

8.2.8 Asbestos Abatement

Introduction

Although this section discusses and provides cost data for the major steps in an asbestos abatement operation, each work location is unique. Each asbestos abatement project is governed by the existing state regulations and the specific requirements of the contract that the owner is using. Caution should be taken to obtain site-specific criteria insofar as possible.

Prior to doing an estimate, the appropriate state agency should be contacted to obtain the following:

• A list of licensed contractors
• A list of approved asbestos landfills
• A copy of the state regulations

This information is usually available and will be mailed at no charge. The contractors' list contains all licensed companies that can bid the job. The list of approved landfills contains the state locations for asbestos disposal (note that the contractor could carry the asbestos into another state). This allows the estimator to determine the haul distance, and the landfill operator can provide actual disposal charges. Landfill operators set their own disposal costs and wide variations in prices are typical.

State regulations have a strong impact on abatement costs. Some states require notification prior to starting the job, a description of the abatement to be done, and the procedures to be followed. Other states require that a state inspector approve the containment prior to abatement and/or another inspector visit the site to approve the job at completion. Waiting for the inspector's arrival and approval of the work adds 4 to 8 hours of incremental crew time. State asbestos regulations are becoming stricter and should be reviewed at least semi-annually.

The owner frequently has specific requirements that will be more expensive than the legal demands. These requirements usually will be defined in the contract documents. The fear of legal liability should a worker or the public be exposed to asbestos is one motivator of these requirements. If the abatement is to occur in an operating facility, the abatement contractor may be given restricted access to the facility so he will not interfere with ongoing operations.

Types of Asbestos-Containing Materials

The problems associated with asbestos are significant. The EPA has surveyed a number of buildings in the United States and has estimated that 20%, or approximately 733,000 buildings, contain friable asbestos-containing materials (ACM).

In order for an estimator to develop cost estimates for asbestos abatement, it is important for that person to understand where asbestos is used. This section discusses:

- What is asbestos?
- How is asbestos used in buildings?
    Surfacing materials
    Thermal system insulation
    Miscellaneous materials
- Friable versus nonfriable ACM
- Inspection of a building for ACM
- Evaluation of inspection results

What Is Asbestos? Asbestos is a naturally occurring mineral. It is distinguishable from other minerals by the fact that its crystals form into long thin fibers. Asbestos minerals are divided into two groups: serpentine and amphibole. The distinction between the two groups is based on crystalline structure. Crysotile is the only mineral in the serpentine group. It is the most common type of asbestos used and accounts for approximately 95% of the asbestos found in the United States. Chrysotile is commonly known as *white asbestos.* There are five types of asbestos in the amphibole group: amosite, crocidolite, anthophylite, tremolite, and actinolite. Amosite is the second most likely type of asbestos found in buildings and is often referred to as *brown asbestos.* Crocidolite, or *blue asbestos*, is also an amphibole used in high-temperature insulation. The three remaining types of asbestos in the amphibole group, anthrophylite, tremolite, and actinolite, are extremely rare and have little commercial value.

The presence of asbestos in buildings does not mean that the health of building occupants is in danger. If asbestos-containing material remains in good condition and is unlikely to be disturbed, exposure will be negligible. However, when asbestos-containing materials are damaged or disturbed, fibers can be released, become airborne, and create a potential health hazard to the building's occupants.

The Uses of Asbestos in Buildings: Asbestos has been used in literally hundreds of products. Collectively these are referred to as *asbestos-containing materials* (ACM). Asbestos was widely used in the past because it is plentiful, low in cost, and has unique properties. It does not burn, is strong, conducts heat and electricity poorly, and is impervious to chemical corrosion. It is well suited for many uses in the construction trades. The uses of ACM in buildings can be divided into three distinct groups: (1) The spray or trowel applied materials used for surfacing; (2) insulation around pipes, boilers, and ducts; and (3) miscellaneous forms, such as cement products, acoustical plasters, textiles, wallboard, ceiling tile, vinyl floor tiles, thermal insulation, and many other materials. Each of these groups can be further divided into two categories: friable and nonfriable. The three groups defined above represent the traditional divisions for building materials. The first two represent mostly friable materials; the third represents mostly nonfriable materials. Asbestos is also used in other materials and items that are not discussed in this section, e.g., hair dryers, toasters, brake shoes, clutch pads, etc.

One of the most common uses for asbestos as a building material was for fireproofing. Fireproofing was commonly sprayed on steel beams used in the construction of multistory buildings to prevent the structural members from warping or collapsing in the event of a fire. Chrysotile was the commonly used asbestos constituent in sprayed-on fireproofing and comprised 5% to 95% of the fireproofing mixture. The balance of the mixture was composed of materials such as vermiculite, sand, cellulose fibers, gypsum, and binder.

Friable Versus Nonfriable ACM: The intent of asbestos regulations is to control the risk of inhalation of asbestos fibers. Because asbestos has been used in many products, the EPA and others chose to separate the asbestos-containing materials into two classes: friable and nonfriable, with friable referring to materials that are easily crushed by hand pressure. It is the friable materials that are of most concern. However, during the demolition or reconstruction of many buildings, the nonfriable materials tend to be broken up and have the potential for releasing asbestos fibers. Therefore, it is necessary to exercise caution in the handlings of these nonfriable materials as well.

Visual inspection is used to determine if a material might contain asbestos; an actual determination can only be made by instrumental analysis. The EPA requires that the asbestos content of suspect materials be determined by collecting a bulk sample and analyzing it by polarized light microscopy (PLM). The PLM technique determines both the percentage and type of asbestos in the bulk material. The EPA also has made the distinction between ACM used inside and outside of buildings. Thus, some of these materials do not have to be inspected and inventoried under the Asbestos Hazardous Emergency Response Act (AHERA) rule (primarily the outside materials such as roofing and vinyl siding). Although the AHERA rules pertain primarily to schools, other Federal facilities may be covered by them in the future. The rules are very stringent and represent the most conservative approach to asbestos abatement.

Surfacing Materials: Surfacing materials can be either friable or nonfriable. Friable forms are either very fibrous and fluffy (sometimes like cotton candy), or granular and cementatious. Because friable materials are more likely to release fibers than nonfriable materials when disturbed, the first priority is to identify the friable surfacing materials that contain asbestos. These materials can be located on walls, ceilings, wide flange beams, or other structural members. The friability of these materials can only be determined by touch. If a powder can be generated by rubbing your hand across it, then the material is friable. However, material that is otherwise friable may be nonfriable if a covering or sealant has been applied to it (e.g., materials that have been painted).

Thermal System Insulation: Asbestos-containing insulation is found on equipment containing hot air or liquid (e.g., pipes, boilers, tanks, and sometimes ducts). These insulation materials may be a chalky mixture of magnesia and asbestos, preformed fibrous asbestos wrapping, asbestos fiber-containing corrugated paper, or any insulating cement. In most cases, the insulating material is covered with a jacket of cloth, paper, metal, or cement. Boiler insulation may consist of thermal blocks or bricks (refractory) or asbestos insulating blankets and is usually covered with a finishing cement. Occasionally, asbestos millboard is used as a stiff outside covering on removable boiler insulation. All block insulation on boilers and breechings, all cements, all pipe fittings, and all gasket materials found should be considered suspect. Frequently ACM is found on insulated pipe fittings. While most of the insulation is ACM free, insulators sometimes use an ACM putty to smooth out the insulation around fittings.

Miscellaneous Materials: Most of the ACM in this category are non-friable such as wallboard, ceiling tile, floor tile, etc. Although not specifically regulated under AHERA, the presence of these materials should be documented and the location included in the permanent records and indicated in the asbestos control program for future considerations. This program should include all the lay-in ceiling tiles and all vinyl asbestos floor tiles. They must either be sampled and analyzed or assumed to contain asbestos. Transite wallboard also should be inventoried and included in the control program, as should fabric material such as stage curtains. Exterior materials such as roofing and vinyl siding need not be identified under AHERA, but should be recorded in the building's permanent record.

Inspection of a Building for ACM: The building inspection involves an investigation of records for the identification of asbestos-containing building materials, an inspection of the building for suspect materials, sampling and analyses of suspect materials for asbestos, and assessing the condition and location of the asbestos-containing building materials and other characteristics of the building. The inspection process consists of the following basic steps:

• Review of the architectural and as-built records
• Inspection of the building for friable materials and other materials that are likely to contain asbestos
• Delineation of homogeneous areas and development of a sampling plan
• Collection of samples and sample analysis
• Collection of information on the physical condition and location of the materials.

The inspector has two major roles: to recognize and sample the asbestos-containing materials in the building, and to assess the building and determine the hazards from the asbestos-containing materials. Both of these are complex functions and are subjective in nature.

Evaluation of Inspection Results: There are several evaluations that can result from the inspection process. The most significant is the hazard assessment. The AHERA rule defines seven hazard ranks, with one (1) being the lowest and seven (7) being the highest. The hazard ranks are based on the condition of the material, potential for contact, influence of vibration, and potential for air erosion. There are other possible hazard evaluation methodologies. Regardless of the evaluation techniques used, it is the hazard rankings that will determine what abatement activities are required. Once the abatement activities are defined, the corresponding cost of those abatement actions can be determined.

Ultimately, all ACM in a building will have to be removed. If not removed earlier, it will have to be removed before the building is demolished. Removing it at demolition is usually cheaper because the building is not occupied and the material removed does not have to be replaced.

Methods of Asbestos Abatement

There are four types of asbestos abatement. They are: removal, encapsulation, enclosure, and management. This section describes two of these methods: removal and encapsulation. Enclosure is the use of traditional construction methods to build barriers around the ACM. Management is a site-specific technique of leaving the ACM in place and operating the facility to minimize health hazards. This section discusses three removal techniques: (1) full containment, (2) glovebag removal, and (3) debris removal. Included is a description of each removal technique, broken down into the following steps: preparation; construction of the decontamination and containment enclosures; asbestos removal; equipment to support the removal, cleanup, and disposal; and restoration. A brief discussion of encapsulation is also included.

The following are several general assumptions that are made for all types of abatement:

• All areas are unoccupied. Before the abatement contractor begins, all personnel, phone systems, communication systems, and movable objects are moved out of the area. In some cases, these items will not be movable and provisions must be made to construct access tunnels to them. The area should be large enough for the work and provide room for the containment, the decontamination enclosure, and an equipment and supplies storage area.

- Electricity, hot and cold water, and sewer access will be provided for the contractor.
- Any object and all interior surfaces that are non-ACM in the containment area are wrapped in two layers of 6-mil plastic.
- All work is based on two 10-hour shifts per day.
- All areas are restored after abatement to a condition equivalent to conditions existing before abatement.
- All work will be done by trained and licensed skilled and unskilled laborers.
- The number of workers and containment size are based on three and one-half workers per 1,000 square feet with a minimum three-man crew, a maximum of 20 workers and a maximum containment size of 15,000 square feet.
- All workers must have physicals, licenses, and respirator training. These are additional costs that a contractor must incur that are not normal under a construction contract and are thus reflected in higher labor or overhead rates.
- An industrial hygienist (IH) technician should be present for abatement activities.

Removal: The following sections discuss the major steps to accomplish asbestos removal. These consist of preparation; decontamination enclosure construction; containment construction; removal, cleanup and disposal of ACM; and restoration. This presentation is based on what is required for a full containment, followed by comments on the differences in methods associated with glovebag removal and debris removal.

Preparation: The first step of preparation is to videotape the area to establish the level to which the area must be restored. This may be omitted, depending on the area renovation required upon completion of the asbestos abatement. Once the videotaping is complete, it is necessary to clean all the fixed objects in the area and to clean and remove all the movable objects from the area. The cleaning is necessary for two reasons: to remove any asbestos fiber contamination on surfaces to be covered with plastic during the abatement and to provide a dust-free environment prior to abatement activities. Post-abatement air samples are first tested to determine the total fiber count in the air. This test costs about $40. If the fiber count is low (less than 0.01 fibers/cc), no additional testing is necessary because the sample would pass the fiber concentration limit even if all the fibers were asbestos. If the fiber count is high (greater than 0.01 fibers/cc), a second test is run to identify the asbestos fibers in the sample. This test costs $300 to $400. The area cleanup prior to abatement reduces the number of the more expensive tests that have to be run.

Next, preparation is made for containment construction. The containment may be constructed in a stepwise fashion and may require some modification of the area in order to gain access for abatement. This can require construction of temporary openings in walls or other temporary structures to gain access to the area. In some cases where abatement is occurring in an area with a high ceiling, scaffolding may be erected and the actual containment constructed on the scaffolding. If demolition is necessary in order to provide access to the containment area, a decision must be made as to whether or not the demolition will potentially release asbestos fibers. If that is the case, then the demolition must be done under containment.

If an area is accessible to the public, a temporary opaque construction barrier between the abatement area and the public is built to secure the area. Special consideration must be given to heating, ventilating, and air conditioning (HVAC) systems used in the area when designing and laying out the abatement area. For example, abatement in an occupied building in which the HVAC system must remain operational requires a coordinated effort to ensure that the HVAC systems can operate safely during the abatement process and that the area where the abatement is occurring can be adequately sealed and contained to prevent asbestos fibers from being drawn into the HVAC system and distributed throughout the building. Likewise, other utilities running through the abatement area can complicate the overall design and layout of the area. For example, emergency shutoff electrical panels and other controls that may be located in the abatement area may require the construction of special tunnels for access during abatement.

If the abatement is to be done above a suspended ceiling, the area above the suspended ceiling must be checked to ascertain if it is being used as a return air plenum. If this is the case and the HVAC system cannot be shut down, a bypass duct will have to be constructed through the plenum area.

A glovebag removal avoids many of the above problems. However, because a glovebag removal can result in a spill or other accident that could release fibers, contingency planning must ensure that there are provisions to isolate the area in the event of a spill.

Debris removal is done without a containment structure. However, all reasonable precautions must be taken to prevent the possible spread of asbestos contamination. Provisions must be made to block off the HVAC system and reduce possible drafts and air movement around the debris to the lowest level. This can include the construction of temporary barriers. The debris to be removed is also moistened to reduce the possibility of asbestos fiber entrainment. Because the debris removal will be done by workers wearing protective clothing and masks, the area should be isolated from the public.

Decontamination Enclosure Construction: The decontamination enclosure is a structure designed to provide an area where asbestos-contaminated equipment can be cleaned off, where workers can shower and remove all asbestos from their bodies, and finally where they can change into appropriate clothing and enter or leave the work area. The decontamination enclosure typically consists of three rooms in series, with the room nearest to the containment area being an equipment room, the next room a shower, and the third room a clean room. The equipment room will have a place to store abatement tools and to stage equipment moving into the area, a place where workers can clean surface debris off their clothing, (perhaps remove tennis shoes or other foot gear that is left in the containment area and used only inside the containment area), and a place to remove and dispose of their work clothing, except for their respirators.

From the equipment room, workers enter the shower room wearing only their respirators. After a thorough shower to remove all asbestos fibers from their bodies and respirators, the workers move into the clean room to dry and put on their normal street clothing. Workers must pass through this system in order to do anything that would require the removal of their respirators, such as use the restroom, eat or drink, take a break, etc. Therefore, it is important that once workers enter the containment area, they are ready to work for a reasonable length of time. An estimate must include allowances for supplies to provide changes of protective clothing and respiratory filters for entrance to and exit from the work area (see Figure 8.6).

For glovebag removal there is no decontamination enclosure. Protective clothing and respirators are worn as a precautionary measure in the event of a bag failure or other accidental release of asbestos fibers.

For debris removal there is no decontamination unit. The debris is sprayed with a liquid to wet down the material and prevent it from becoming airborne. Workers wear protective clothing to pick up and bag

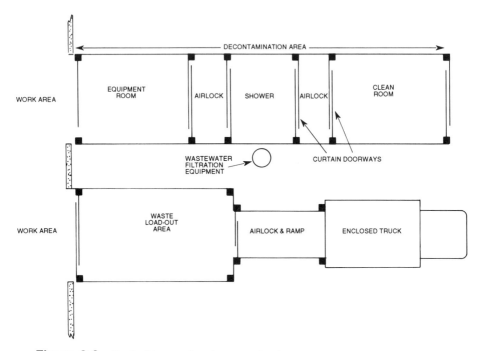

Figure 8.6 Typical layout for decontamination areas.

the debris and clean the surrounding area. Upon completion of the work they decontaminate their suits with a HEPA vacuum, remove them, and exit the area.

Construction of the Containment Enclosure: For conventional removal, the containment enclosure consists of the critical barrier and subsequent layers of plastic. The critical barrier consists of the plastic used to cover and seal all doors, windows, and other openings. The critical barrier uses the walls, floors, and other segments of the area to establish an enclosure. Next, multilayers of polyethylene are used to cover all the exposed surfaces inside the area, which are not being abated. Figures 8.7 and 8.8 represent how this system is assembled.

Before installing the containment enclosure, all surfaces inside the area that do not require abatement must be cleaned and free of asbestos. After the containment enclosure is constructed, the asbestos-negative air ventila-

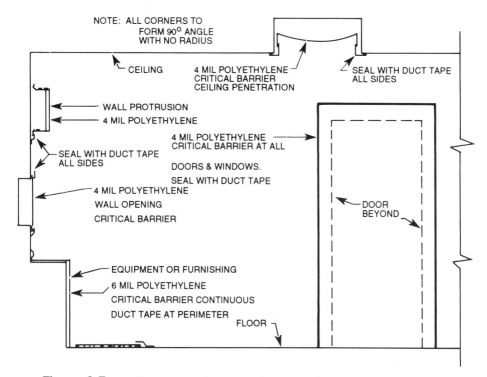

NOTE: ALL CORNERS TO
FORM 90° ANGLE
WITH NO RADIUS

CEILING

4 MIL POLYETHYLENE
CRITICAL BARRIER
CEILING PENETRATION

SEAL WITH DUCT TAPE
ALL SIDES

WALL PROTRUSION

4 MIL POLYETHYLENE

4 MIL POLYETHYLENE
CRITICAL BARRIER AT ALL

SEAL WITH DUCT TAPE
ALL SIDES

DOORS & WINDOWS.

SEAL WITH DUCT TAPE

4 MIL POLYETHYLENE

WALL OPENING

CRITICAL BARRIER

DOOR
BEYOND

EQUIPMENT OR FURNISHING

6 MIL POLYETHYLENE

CRITICAL BARRIER CONTINUOUS

DUCT TAPE AT PERIMETER

FLOOR

Figure 8.7 Sealing features for a containment enclosure.

tion system is installed. The pressure in the containment is maintained below the pressure of the surrounding areas and the air vented from the containment is filtered to remove asbestos fibers. The negative air system should provide approximately four air changes per hour and should be monitored. A reading of 0.02 to 0.04 inches of water differential is adequate to maintain the containment at a negative pressure. For glovebag operations there is no negative air system required. Also, the walls are not covered with plastic sheeting as indicated in Figure 8.7. Critical barriers should be installed, similar to those in Figure 8.6, as appropriate for the size and scope of the glovebag removal project. For debris removal, no negative air system is used and, usually, no critical barrier is installed.

Removal: The first abatement activity of the removal process is the wetting of asbestos with water containing a surfactant or an amending

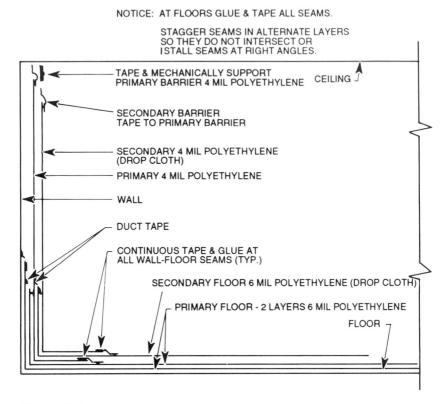

Figure 8.8 Construction features for a containment enclosure.

agent additive. These additives enhance the water's ability to penetrate into the surface of the asbestos and to wet it. Asbestos fibers, once wetted, are more difficult to entrain. The removal then proceeds using tools to cut, scrape, or otherwise remove the ACM. The tools will include putty knives or other types of scrapers, scrub pads, wire brushes, sponges for wiping and cleaning surfaces, rags, and even some power tools. The ACM must be kept wet during the removal process. Consumable protective clothing and respirator filters must be used by the workers during removal. Typically, a worker will be clothed in a protective suit made of a Tyvek material. The suit will have a hood to cover the head and built-in feet or secondary booties to cover the feet. Duct tape is used to secure the suit around the wrists and ankles. The respirators used by the workers must be worn during showering and, consequently, the air filters must be replaced each time a worker goes through the decontamination procedure. For estimating purposes, four clothing changes per worker per shift are typical. This allows the workers to take a break in the morning and afternoon as well as their normal lunch break. Toweling material must be provided so the workers can dry off as they exit the areas.

Other special equipment requirements include the shower facilities. Hot water must be provided. Because the shower water will contain asbestos fibers it is necessary to filter the waste water before disposal. Other items typically used in a containment include the surfactants mixed with the water, warning stickers that must be applied to each of the ACM disposal bags, spray systems for applying the surfactant, buckets for holding rinse water, water used in the decontamination of the area, knives to cut plastic and duct tape, HEPA-filtered vacuum cleaners for the cleanup of debris, and the ventilation fans with HEPA filters to exhaust air from the area. A minimum of two HEPA-filtered ventilation fans must be provided in any abatement area so that there is always a backup.

There are some possible variations for ACM waste removal from the abatement area. The ACM contaminated waste can be placed in plastic bags and sealed. The bags are cleaned, placed in a second bag and sealed, and then removed from the containment for disposal. Some contractors use a large trailer-mounted vacuum system to assist in the removal of the ACM from the containment. These units are located outside the abatement area and have a containment enclosure. The ACM is immediately placed in bags within the trailer or left in a large vacuum container for later bagging.

For glovebag operation, the clothing requirements are the same as for a conventional abatement. However, it is no longer necessary to provide the HEPA-filtered ventilation units. Glovebag removal operations typically use fewer consumable materials to accomplish the work.

For debris removal, consumable supplies are the same but fewer than those for conventional containment; however, HEPA-filtered ventilation units are not required.

Clean-Up and Disposal of ACM: Once the asbestos abatement begins, all equipment, clothing, and anything else used inside the removal area must either be decontaminated before removal from the containment area or be discarded as asbestos-containing waste. Consequently, few power tools are used inside an abatement area, since those tools are virtually impossible to completely decontaminate. When power tools such as HEPA-filtered vacuums are used, those units are decontaminated, sealed in plastic, and saved to be used at other work sites.

All items not decontaminated will be double-bagged and placed in disposal drums. The drummed material is removed from the abatement area and transported immediately to an approved disposal facility or placed in a disposal box or area. Workers must wear protective clothing and respirators whenever they handle the waste containers. Many states require that the asbestos-containing waste be transported only by licensed haulers. All states require that the material be disposed of at an approved landfill. Approved landfills require that the material be unloaded and buried immediately, that the operator of the equipment putting the earth cover over the material wear a respirator and protective clothing, that the workers unloading the materials from the transport vehicle wear protective clothing and respirators, and that no visible dust is caused by the process. Disposal regulations allow municipal landfills to be used for asbestos disposal. However, not all municipal landfills care to institute the necessary procedures to handle asbestos-containing waste. Thus, transportation to the disposal site can be expensive. Special contract requirements for asbestos disposal can add to those costs. Disposal costs for small jobs can be inordinately high because there typically are minimum transport fees and disposal costs.

Cleanup operations require that all surfaces inside a containment be wet-wiped and cleaned prior to the completion of the removal project. For glovebag and debris removal, the area is not under containment. Therefore, there are no requirements to wet-wipe surfaces in the area where the removal occurs.

There are special requirements for air sampling and inspection before and after the removal process. These requirements are discussed in the section on Support Services.

Restoration: The aim of restoration is to return the abated area to its original condition. Restoration requirements and costs will vary based on the complexity of the removal project. Because restoration takes place after barriers have been removed, conventional estimating practices are appropriate.

One of the areas that is sometimes difficult to estimate and even more difficult to accomplish, is the restoration of abated surfaces. It is common for the abatement contractors to use structural foam materials to block off openings and to make extensive use of spray adhesives and duct tape to secure plastic to the walls, floors, ceiling, and other fixtures. The cost of removing the foam materials and restoring the surfaces on which duct tape and spray adhesives were used must be factored into the total restoration cost.

It may be necessary for the abatement contractor to paint walls; repair holes in walls or other surfaces; rehang doors; and reinstall light fixtures, switchplate covers, ventilation system ductwork and grills, etc. Lay-in ceilings are frequently replaced with new materials and the old materials discarded. If the asbestos-containing materials removed during the process were fireproofed, it is necessary to refireproof the appropriate structures with non-asbestos-containing materials. Frequently, this refireproofing process results in overspray and additional cleanup costs.

Encapsulation: Encapsulation may be the only practical abatement technique or the most cost effective for some situations. Encapsulation does not remove the ACM from the building, and it is not a permanent solution because the ACM must be removed prior to demolition. Additionally, the encapsulated ACM potentially could be disturbed causing asbestos fibers to become airborne.

Encapsulation uses a material that penetrates into the surface of the ACM and binds it into a more rigid and solid mass. Encapsulation reduces the possibility of asbestos fibers becoming airborne. No containment or negative ventilation system is required. The process consits of videotaping the area, protecting surfaces in the area that will not be encapsulated, spray-applying the encapsulant material (like paint), cleaning the overspray and area, and inspection and acceptance of the area. Typically, air samples will be collected by the building owner's industrial hygienist to ensure that all surfaces are sprayed with the appropriate quantities of the encapsulating material.

Encapsulation also is done as the last step in a removal project. Typically, encapsulant is sprayed on all the abated surfaces as a final precaution to lock down fibers that might not have been removed.

Support Services

Asbestos abatement requires specialized support services, including support from the building owner and occupants. At a minimum, an abatement project needs an industrial hygienist, laboratory services to analyze the samples, and a construction manager knowledgeable in asbestos abatement activities.

Building Owner Support: The building owner must provide support for the asbestos abatement project. This support includes free access to the area in which the asbestos is located; adequate space outside the abatement work area for storage of necessary supplies and of the asbestos-containing waste material; and knowledgeable information on the HVAC, mechanical, electrical, and communication systems servicing the abatement area. Obviously, asbestos abatement in an unoccupied building is less complicated than in an occupied building.

When the building is occupied, it is essential for the building owner and abatement contractor to work together for a successful abatement project. Special considerations are needed for entry and egress of abatement and building workers and for the appropriate notification of building occupants. The industrial hygiene responsibilities become much more critical for occupied buildings because a breach in the containments will not only contaminate the building, but will also expose building occupants to asbestos.

Industrial Hygiene Services: The asbestos removal techniques discussed require that the removal be performed according to various local, state, and Federal regulations. The removal contractor must comply with OSHA regulations regarding worker exposure to asbestos and must provide industrial hygiene services during the work to satisfy OSHA. Additionally, the building owner will provide industrial hygiene services to ensure that the building owner's interests are protected during the asbestos abatement work. The owner's industrial hygiene representative, through air sampling and inspection, is responsible for verifying that the abatement contractor performs his work according to the contract documents, protects the abatement workers, and represents the building owner's interests.

The industrial hygienist (IH) collects background air samples prior to abatement to establish a clearance baseline. Once the containment barrier is installed, but prior to actual abatement, the industrial hygienist will inspect the area to ensure that all the surfaces are clean and to verify the integrity of the containment barrier (pre-abatement inspection). During the removal process, the owner's industrial hygienist will collect area air samples to demonstrate that no contamination is occurring. The IH also ensures that abatement workers wear proper respirators and protective clothing and follow proper decontamination procedures. Daily reports and sample logs are prepared by the IH to document the abatement project. Upon completion of the removal and after the cleanup, the owner's industrial hygienist will inspect the area and conduct clearance sampling (post-abatement inspection).

Clearance air samples are obtained using an aggressive air sampling technique where air is blown (a leaf blower can be used) around the abatement area to stir up any fibers. Once the building owner's industrial hygienist accepts the area as free of asbestos-containing materials, the contractor will remove the critical barriers and restore the area. Upon completion of restoration, the owner will inspect the area.

The air monitoring is done for two reasons: to locate a problem occurring during the abatement so it can be corrected and to document that the abatement was done properly. Persons around the abatement project who think that they have been exposed to asbestos could file a lawsuit against the contractor or owner. Also, anyone around the project who develops an asbestos related disease 20 or 30 years later could have basis for a lawsuit. For this reason, many owners and contractors feel they need to collect their own air-monitoring data and store that data long-term. In addition to the written reports, many owners catalog and save the filters from the air samplers.

These procedures are basically the same for all the removal situations discussed. The number of air and bulk samples collected will be modified as appropriate to match the conditions of each technique. These modifications do not appreciably affect the industrial hygiene costs for the abatement.

Laboratory Services: Laboratory services are used to analyze the air and bulk samples obtained by the IH. Most air samples are analyzed using the phase contrast microscopy (PCM) technique to count the fibers. However, if the fiber count is too high (greater than 0.01 fibers/cc), the transmission electron microscopy (TEM) technique is used to differentiate the asbestos fibers from other fibers. TEM analysis costs much more than PCM ($400 versus $40). Therefore, as an important cost savings measure, every reasonable effort should be made to keep the fiber count down (by extensive cleaning of the area prior to abatement) and to use PCM instead of TEM analysis. However, there are some sampling situations that require TEM; for example, final clearance samples for AHERA abatement.

Abatement typically is performed 24 hours per day. Therefore, laboratory services may be necessary at any time of the day. Also, laboratory records should be kept for 30 years for liability reasons. Both of these requirements increase the cost of laboratory services.

Asbestos Abatement Project Cost Estimating

This section presents the components of asbestos abatement estimates and cost estimating methods, formulas, and unit costs.

Components of Asbestos Abatement Estimates: The following components should be included in an asbestos abatement estimate:

- Direct costs: building inspection; preparation for asbestos removal; asbestos abatement; decontamination and cleanup; transportation and disposal; restoration.
- Other contractor costs: general conditions; contractor labor burden, overhead, and profit; asbestos liability insurance; bonds.
- Third-party or subcontractor costs: industrial hygiene services; laboratory services; construction management (optional).

The total of the direct, contractor, other, and third-party costs forms the estimate of the project's cost. The accuracy of the estimate will depend on the specific knowledge of the job conditions and the scope of work (SOW). Preliminary estimates typically have a ±40% accuracy; estimates performed after preparation of the SOW will be ±20% or better.

Order-of-Magnitude Estimates: Order-of-magnitude estimates generally do not include adjustments for the specific site conditions, labor productivity, or crew sizes. As with any estimate, the estimator must fully document the basis of quotations, assumptions, and any items specifically included or excluded. The factors shown in Table 8.27 can be used for order of magnitude estimating purposes.

Definitive Estimates: Definitive asbestos abatement estimates should have an accuracy range of ± 20%. This type of estimate can be developed

Table 8.27   Order of Magnitude Asbestos Abatement Cost Estimate Multipliers

---

Ceiling Insulation (Under Containment):

   For 500 - 5,000 SF of Ceiling Area Use $37 to $21 per SF
   For 5,001 - 20,000 SF of Ceiling Area Use $21 to $13 per SF

Pipe Insulation (Under Containment):

   For 100 - 1,000 LF of 6" Pipe Use $131 to $51 per LF
   For 1,001 - 3,000 LF of 6" Pipe Use $51 to $45 per LF

Pipe Insulation (Glovebag Technique):

   For 100 - 1,000 LF of 6" Pipe Use $75 to $70 per LF
   For 1,001 - 3,000 LF of 6" Pipe Use $51 to $45 per LF

---

when key project data (i.e., specifications, labor wage rates, quantities) are known. Good definitive estimates include allowances for items such as site conditions, labor productivity, specific restoration materials, etc.

Preparation for asbestos removal includes making a video of the abatement area, precleaning the abatement area, erecting temporary personnel and decontamination structures and/or enclosures, covering fixed objects with plastic, and installing the containment ventilation units. Asbestos abatement includes all the encapsulation or removing of ACM from either architectural or mechanical building components.

Decontamination and cleanup includes the specialized cleaning and dismantling of temporary structures and enclosures as well as sealing abated surfaces with chemicals. It also includes the cost of small tools and consumable supplies, HEPA filtration equipment, personnel respiratory protection, and disposable clothing. Disposal costs include the loading, transportation, and disposal of asbestos debris at EPA-approved landfills. Restoration costs include the replacement of contaminated materials. Other contractor costs include general conditions, labor burden (including contractor IH services), the abatement contractor's project overhead and profit, and specialized bonds and insurance.

Project management and industrial hygiene services include the construction management costs, the sampling and analytical costs, and the owner IH cost. Also included are technical, professional, and certified industrial hygiene project support.

Cost Estimating Methods, Formulas, and Unit Costs

This section can be used to estimate the owner's cost of an asbestos abatement project. The following section discusses other possible cost considerations to the owner. Each section describes what is included in each cost estimating formula. These descriptions are followed by a tabular presentation of the formulas.

Building Inspection: The building inspection and sampling requires approximately one day for every 50,000 to 100,000 square feet of building floor area (not including sample analysis), depending upon the building complexity, with an average of approximately 90,000 square feet per man day. A comprehensive building inspection report requires approximately 3 days to complete (based on a 500,000 square foot facility). Samples are analyzed at roughly $30 each with a minimum of ten samples per building, a range of one to five samples per 3,000 square feet of building. The cost of the survey ranges from $37 to $52 per manhour with an average of $42 per manhour. On the basis of these averages the cost of the building inspection can range between 2 cents and 7 cents per square foot, depend-

Table 8.28 Building Inspection Costs in Dollars per 1,000 Square Feet

	Low	Average	High
Inspection	$2.93	$3.72	$8.37
Sample Analysis	$12.21	$20.51	$61.01
Report Writing	$1.00	$2.01	$4.02
TOTAL	$16.14	$36.24	$73.40

ing on the analysis technique chosen. See Table 8.28 for a tabular presentation of the data.

Preparation for Asbestos Removal: Preparation activity starts with the abatement contractor gaining access to the area and ends with the pre-abatement inspection by the IH immediately prior to the start of asbestos removal. This section discusses the steps the contractor follows and the associated costs. The next sections discuss each step in the preparation activity, and the costs are given in Table 8.29. All costs are mid-1991 dollars for the Dallas, Texas, area.

Table 8.29 The Cost of Preparation for Asbestos Removal Activities (Total Cost = L × SLR + M + E)

Description	Quantity	Unit of Measure	Labor (L) Hours	Material (M) $	Equipment (E) $
PRE1 Videotape	(CA/5000)+2	HR	QTY*2	QTY*5	QTY*10
PRE2 Exterior barrier walls	Wall area	SF	QTY*0.0324	QTY*0.325	N/A
PRE4 Public access tunnels	Tunnel length	LF	QTY *1.04	QTY*15	N/A
PRE5 Scaffold open area	of scaffold	SF	QTY*.035	N/A	QTY*2
PRE6 Scaffold congested area	of scaffold	SF	QTY*0.07	N/A	QTY*1.25
PRE7 Rolling scaffold	Per 4′ x 6′ unit	EA	N/A	N/A	QTY*100
PRE8 Plastic on objects	AFO	SF	QTY*0.05	QTY*0.10	N/A
PRE9 Decon Unit	[(CA)*3.5/1000]/5	EA	QTY*40	ATY*418	N/A
PRE10 Containment walls	(CP)*(CH)	SF	QTY*-0.02	QTY*0.05	N/A
PRE11 Containment floors	CA	SF	QTY* 0.05	QTY*0.10	N/A
PRE12 Containment tunnel	Length of tunnel	LF	QTY*0.45	QTY*7.00	N/A

Definitions:

CA Containment area
CP Containment perimeter
CH Containment height
AFO Area of fixed objects
SLR Standard labor rate

Video: Once the area has been turned over to the abatement contractor, the area's condition is documented. The contractor, accompanied by the building owner's construction manager, prepares an audio-video tape with a listing of possible problem areas. If the area is to be restored, the video records architectural as well as insulation details. Video requires a minimum of 2 hours including travel, set up, and playback time. The actual video recording requires about 1 hour per 5,000 square feet. In some cases lights have to be used. Video costs range from $31 to $52 per hour (see Formula PRE1 in Table 8.29).

Public Barriers: After videotaping the area, the next step is to isolate it. Depending on the layout of the area, the abatement contractor may need to construct public barriers and access tunnels. Exterior barrier walls are generally plywood and 2 × 4 walls built to form the containment area. Interior public barriers are built to isolate the public from the abatement work and usually form a wall of the containment. Because the interior walls are exposed to the public they are usually made of a more attractive material such as vinyl-coated sheetrock or paneling, and thus cost more. The cost of these walls are given in equations PRE2 and PRE3 in Table 8.29. When these walls form part of the containment they are coated with plastic on the side with the studs. The cost of the plastic is given in equation PRE10.

Access tunnels are to allow public passage through an area being abated. The tunnels shown in PRE4 are vinyl-coated sheetrock on a 2 × 4 framework and are 8 feet high and 2½ feet wide with flush mount lights. The cost equations include the cost of installation and removal. The cost assumes that the material (other than the plastic) is used three times.

Containment: After installation of the barrier walls and public access tunnels, the abatement contractor constructs the containment. The erection of scaffolding, the containment and decontamination unit, and the wrapping of nonmovable objects in plastic usually occur simultaneously.

- For scaffolding, see Formulas PRE5 through PRE7 in Table 8.29. The total costs include all labor, material, and equipment necessary to bring the scaffolding to and from the area, its erection, tear down, and any rental fee.
- For plasticizing the objects, containment walls and floors, see Formulas PRE8, PRE10, and PRE11 in Table 8.29. The total costs include the cost of precleaning the critical barrier, installing two layers of plastic, wet wiping, removing, and placing the material in disposal bags at the end of the job. For a plastic access tunnel, see Formula PRE12.

The total costs include the studs and two layers of plastic, which are installed, wet-wiped, removed, and placed in disposal bags at the end of the job.

- For a decontamination unit, see Formula PRE9 in Table 8.29. Each decontamination unit includes a 100-square foot clean room, a shower unit with a hot water heater and a dirty water filter that filters out any asbestos fibers before the water is sewered, and a 100 square foot dirty or equipment room. The total cost includes studs and plastic construction, wet wiping, and removal and placement of the contaminated plastic in disposal bags at the end of the job. Each unit will accommodate five people. The cost equation assumes that 3.5 people will be used per 1,000 square feet of containment.
- The containment walls and floors shown in equations PRE10 and PRE11 assume that two layers of plastic are used. The containment tunnel is built of 2 × 4s with plastic walls and a plywood ceiling. The plywood ceiling is used as a support deck if abatement workers have to work on top of the ceiling.

Asbestos Abatement: This section discusses the cost of asbestos abatement using two generally accepted methods: removal of the asbestos-containing material or its encapsulation. Unit manhour and material rates for various architectural and mechanical components that could be encountered during a project are provided.

Removal: Labor costs comprise most of the total removal cost. Thus, all items that affect a worker's ability to work influence the removal cost. These include but are not limited to such things as ambient temperatures, condition of ACM, accessibility, level of worker protection, worker experience, and quantity of ACM. Cost formulas have been developed for three categories of removal:

- Within a containment structure
- Glovebag removal
- Debris removal

The above three categories are discussed as they apply to friable ACM. The removal of nonfriable ACM such as floor tiles or exterior panels is substantially less expensive than the cost of removing friable ACM. However, nonfriable ACM removal does require careful handling of the building components so that they are not broken during the process, thereby causing a fiber release. The contaminated materials are wrapped in plastic for disposal. Workers dress in protective clothing with breathing equipment; however, a containment structure is usually not required. Table 8.30 is a reference table that illustrates labor hours required for nonfriable ACM removal.

Friable ACM removal is the most common type of removal. Two methods of removal are discussed; containment (full-size and mini) and glovebag removal. The direct labor hours for different removal situations are shown in Tables 8.31–34. Note that these tables do not include the cost of building the containment, monitoring, protective equipment, or any costs other than the actual hours required for removal.

Table 8.31 gives labor hour factors for removal of ACM from mechanical surfaces, pipe, and pipe fittings under containment conditions. Table 8.32 gives similar information for the removal of ACM from structural steel beams. Table 8.33 shows the hours required for glovebag removal from pipe and pipe fittings and also includes the price of the glovebag. No containment is required for glovebag removal, but the workers must wear protective clothing. Table 8.34 provides information for crawl space abatement. The factors for crawl space abatement should be used when the area has a vertical clearance of 3 feet or less.

Table 8.35 gives abatement equipment costs. The HEPA ventilation units are based on the size of the containment with a minimum of two per area. The number of units is determined by computing the volume of the containment in cubic feet and dividing by 30,000. If the number is more than 2, round it to the next larger whole number.

Table 8.30 Labor Hours for Removal of Nonfriable ACM

Item	Unit of Measure	Labor Rate
Transite Thermal Barrier	SF	110% of Installation labor
Exterior Transite Siding	SF	110% of Installation labor
Floor Tile	SF	87% of Installation labor
Transite Duct Liners	LF	25% of Installation labor
Transite Hood Liners	LF	25% of Installation labor
Transite Counter Tops	LF	140% of Installation labor

NOTE: Work can be performed by unskilled workers.

Labor rates are from the report "Demolition Cost Estimating Study," 1985, by PEI Associates, Inc. The demolition labor shown here is increased 50 percent over non-ACM demolition to cover the decreased productivity caused by protective clothing and the required careful handling.

Table 8.31 Labor Hours per Unit of Measure (UOM) Required to Remove ACM Under Containment (Removal Only)

Mechanical Component	Unit of Measure	Hours/Unit of Measure
Boiler surfaces	SF	0.10
Boiler breeching	SF	0.10
Tanks	SF	0.10
Duct insulation	SF	0.11
Piping		
1/2"	LF	0.014
3/4"	LF	0.017
1"	LF	0.022
2"	LF	0.039
3"	LF	0.057
4"	LF	0.074
6"	LF	0.11
8"	LF	0.14
10"	LF	0.18
12"	LF	0.21
Fittings		
1/2"	EA	0.041
3/4"	EA	0.052
1"	EA	0.065
2"	EA	0.12
3"	EA	0.17
4"	EA	0.22
6"	EA	0.33
8"	EA	0.42
10"	EA	0.53
12"	EA	0.63

The number of sets of disposable clothing is computed by allowing for four changes per 10-hour shift. A set of two respiratory cartridges is required for each change of clothing. This protective equipment must be worn from the time that area preparations begin until the area is declared clean. These costs are shown in Table 8.35.

Encapsulation: Encapsulation refers to the spraying of ACM with a sealant. It should only be used on granular, cementatious material, commonly known as acoustical plaster. Material that is delaminated or deterior-

Table 8.32 Labor Hours per Unit of Measure (UOM) Required to Remove ACM From Structural Steel Beams (Removal Only) in a Containment

Beam Size	Unit of Measure	Hours/Unit of Measure
W8X12	LF	0.20
W8X17	LF	0.26
W10X19	LF	0.27
W10X31	LF	0.33
W10X49	LF	0.41
W12X16.5	LF	0.29
W12X40	LF	0.39
W12X58	LF	0.44
W14X26	LF	0.35
W16X31	LF	0.39
W18X45	LF	0.47
W24X55	LF	0.56
W24X130	LF	0.74
W30X108	LF	0.74
W36X150	LF	0.88
Columns		
W6X12	LF	0.17
W8X17	LF	0.22
W8X31	LF	0.28
W10X19	LF	0.23
W10X49	LF	0.35
W12X16.5	LF	0.25
W12X40	LF	0.33
W12X58	LF	0.37
W14X26	LF	0.30
W16X31	LF	0.33
W18X45	LF	0.40
W24X55	LF	0.48
W24X130	LF	0.63
W30X108	LF	0.63
W36X150	LF	0.75
Overspray	SF	0.37
Architectural Surface	Unit of Measure	Hours/Unit of Measure
Corrugated or fluted	SF	0.11
Smooth (steel)	SF	0.10
Smooth concrete or plaster	SF	0.13
Rough plaster	SF	0.15
Concrete surfaces		50% more than equivalent steel surfaces

Table 8.33 Labor Hours per Unit of Measure (UOM) Required for Glovebag Removal (Includes the Cost of the Glovebag)

Piping Size	Unit of Measure	Hours/Unit of Measure
1/2"	LF	0.30
3/4"	LF	0.35
1"	LF	0.40
2"	LF	0.50
3"	LF	0.55
4"	LF	0.69
6"	LF	0.70
8"	LF	0.80
10"	LF	0.90
12"	LF	0.95
Fittings		
1/2"	EA	0.90
3/4"	EA	1.05
1"	EA	1.20
2"	EA	1.50
3"	EA	1.65
4"	EA	2.07
6"	EA	2.10
8"	EA	2.40
10"	EA	2.70
12"	EA	2.85

Table 8.34 Labor Hours per Unit of Measure (UOM) for Crawlspace[a] Decontamination

Activity	Unit of Measure	Hours/Unit of Measure
Preclean floors and walls	SF	0.16
Cover surfaces with plastic	SF	0.17
Remove asbestos material		
Piping		
1"	LF	0.43
1 1/2"	LF	0.48
2"	LF	0.50
3"	LF	0.55
4"	LF	0.60
6"	LF	0.75
8"	LF	0.86
Ductwork	SF	1.02
Fittings	EA	1.65
Remove plastic	SF	0.02

[a] A crawlspace is defined as an area with less than three feet of height.

Table 8.35 Abatement Equipment Costs

Mechanical Components	UOM	Labor Productivity Factor, hrs/UOM	Material Unit Cost, $/UOM	Equipment Unit Cost, $/UOM
Negative air equipment HEPA filters	Days	0	26.98	25.81
HEPA vacuum cleaner and maintenance	Days	0.26	9.21	12.33
Personnel Protection Cartridges (set of 2)	EA	0	3.14	0
Full body disposable clothing	EA	0	3.60	0

ated should not be encapsulated because material in this condition may be pulled down or blown off by the sealant application.

Two types of sealant exist for encapsulation. A penetrating sealant penetrates the ACM, adheres to the substrate by binding together with the asbestos fibers and other material components, and offers resistance to damage from impact. Penetrating sealant costs approximately $7.32 per gallon and between 2 to 10 gallons will cover 25 square feet depending on the thickness applied. A rule of thumb for coverage is no more than 10 square feet per gallon with a continuous, unbroken coating. Bridging sealant forms a tough skin over the ACM to protect it from damage. Bridging sealant typically is more expensive, costing approximately $19.61 per gallon, with coverage of 3.2 gallons per 100 square feet.

Both types of sealants may be applied with airless spray equipment. One recommended method is to apply a light mist coat, then apply a full coat at a 90-degree angle to the direction of the first. An airless sprayer costs approximately $26.15 per day. No containment structure is necessary for encapsulation. However, areas that are not to be encapsulated must be covered with plastic and personnel must wear protective clothing and respirators.

Average labor, material, and equipment costs for encapsulation are shown in Table 8.36. The costs for the plastic coverning include the installation of one layer of plastic to catch overspray, and removing and placing the overspray and plastic in a disposal bag at the end of the job.

Debris Removal: Debris removal can be performed with a minimum containment structure. The removal of asbestos debris from floor surfaces consists of wetting the material to minimize fiber release when picking up and bagging large pieces, and then HEPA-vacuuming the area. If the ACM

Table 8.36 Asbestos Encapsulation Costs (Also Can be Used for Lockdown Application)

Description	UOM	Labor (L) hrs/UOM	Material (M) $/UOM	Equipment (E) $/UOM
Penetrating				
Ceilings and Walls	SF	0.04	0.29	0.01
Columns and Beams	SF	0.07	0.29	0.03
Pipes (12" diameter or less)	LF	0.17	0.38	0.07
Plastic covering	SF	0.05	0.05	N/A
Bridging				
Ceilings and Walls	SF	0.04	0.63	0.01
Columns and Beams	SF	0.07	0.63	0.03
Pipes (12" diameter or less)	LF	0.17	0.81	0.07
Plastic covering	SF	0.05	0.05	N/A

is on the ground, it may also be necessary to remove the top 3 to 5 inches of soil if that soil has been contaminated. Workers wear disposable clothing and breathing equipment. The debris is disposed of as asbestos-containing waste. A unit labor rate of 0.010 hours per square foot should be adequate for estimating purposes.

Decontamination and Cleanup: As part of the removal cost, the workers will bag the removed material and wet-wipe the area until it is clean and free of visible material. In many cases the abated area is sprayed with an encapsulant to lock down any fibers that might remain prior to final air clearance sampling. The costs for encapsulation shown in Table 8.36 may also be used to estimate lockdown applications. Once the final cleaning (wet wiping) and air sampling is successfully completed, the containment barriers are removed and the area is prepared for restoration.

Transportation and Disposal: The labor to bag the ACM waste is included in the abatement activities. The bagged ACM is temporarily stored inside the containment. All costs incurred for removing the bags of ACM from the containment and disposing of them, together with the material price of the bags and ties, are included in transportation and disposal. For large abatement projects, a central storage area may be established that adds to the handling, but reduces transportation costs overall by allowing for accumulations of large loads. This is accounted for in the cost components for transportation and disposal.

The number of disposal bags and drums required will depend on the type(s) and amount of ACM removed. Each bag will hold approximately 1.57 cubic feet of asbestos fireproofing. Smaller amounts of pipe insulation and other shaped forms of ACM will fit into the bags. The bags and ties cost $0.63 each. As they are removed from the containment, they must be cleaned and placed in either a second bag or in a fiber drum. In some cases both a second bag and a fiber drum are used.

The second bag or drum is sealed as soon as it is filled. Each bag requires approximately 0.05 labor hours to remove it from the containment. A fiber drum costs $12.55. Each drum will hold from two to three bags, depending on the type of ACM and the size of the drum.

The drums of ACM are loaded into a closed top truck or dumpster and hauled to an approved dumpsite. ACM waste must be sent to an approved landfill, not a hazardous waste landfill. Transportation regulations vary from state to state and can significantly affect the cost. Some states require ACM waste to be handled by licensed haulers. Typically, the hauling and dumping is subcontracted.

The cost of hauling the drums to the dumpsite, unloading, and burying them is estimated at $5.23 per drum, or approximately $29.98 per cubic yard. The dumping fees are estimated at $1.57 per drum, or $5.99 per cubic yard. At the dumpsite, the drums are unloaded in the landfill and buried. Most landfills will bury them immediately. The dumping cost per drum or per cubic foot will depend on the quantity of waste to be disposed of. It will also vary depending on the state regulations.

Table 8.37 summarizes the transportation and disposal costs of asbestos abatement projects.

Restoration: Restoration (or sprayback) includes the cost of replacing the abated materials with non-asbestos-containing materials. The estimator should generally assume that the replacement material will have the same acoustical, thermal, and/or insulating properties as what was originally provided; however, if project specifications are available they must be consulted. Standard estimating manuals and guides can be used to estimate restoration costs.

Contractor Other Costs: These are costs incurred to support the overall project effort. The major indirect costs are general conditions, contractor labor burden, liability insurance, bonds, overhead, and profit. Some state legislatures are considering imposing a services tax, similar to a sales tax.

Table 8.37 Summary of Transportation and Disposal Costs

Activity	Cost	Cost/Cubic Yard[1]	Comments
Bags & Ties	$0.63 each	$10.80	Each bag holds about 1.57 cubic feet of ACM.
Removing Bags from Containment	0.05 hours each	$11.72[2]	The bags are placed in a second bag, or a fiber drum or both.
Fiber Drums	$12.55 each	$23.99[3]	Each drum will hold three bags.
Hauling and Dumping	$5.23 per drum	$29.98	Hauling drums to dump site and dumping.
Dumpsite fees	$1.57 per drum	$5.99	Fees for dumping at dumpsite.
Total Cost/Cubic Yard		$82.48	

[1] Cost per Cubic Yard assumes 3 bags per fiber drum

[2] Assumes Average Labor Rate of $13.63/hour for Dallas, Texas

[3] Assumes reusing each drum three times.

General conditions are constractor costs for items that are unique to the job. The general conditions can be computed two ways. The first approach is a line-by-line listing and costing of the components to be included under general conditions. The second is to use a percentage of the total direct labor calculated for the project. Traditional estimating techniques can be used to develop these costs. The factor ranges from 4% to 8% of the estimated labor cost, with an average of 6%.

Contractor labor burden is the line item cost for the abatement workers' social security, workmen's compensation, and company benefits. Traditional estimating techniques can be used to develop these costs. The percentage factor typically ranges from 17% to 24%, with an average of 20% of direct labor cost dependent on the state and/or the company.

Asbestos liability insurance is the specialty insurance sometimes required by the contractor or owner to protect against legal claims. The insurance currently costs between 10% and 15% of the project's cost. Bonds required as a condition of the contract should be considered an other cost. They can range in cost from 5% to 10% of a project's costs, with an average of 6%. The abatement contractor's overhead typically covers costs for such items as office trailers, telephones, secretarial services, office supplies, and utilities. Overhead ranges from 20% to 30%

Table 8.38 Contractor Other Cost Components

Cost Component	Percentage			Percentage Based On
	Low	Average	High	
General Conditions	4	6	8	Direct labor cost
Labor Burden	17	20	24	Direct labor cost
Asbestos Liability Insurance	10	12	15	Total estimated cost
Bonds	5	6	10	Total estimated cost
Overhead	20	25	30	Total estimated cost
Profit	5	8	10	Total estimated cost

of the project's costs. Profit can range from 5% to 10% of the project's costs. Refer to Table 8.38 for a summary of these components.

Industrial Hygiene Services: Industrial hygiene services are third-party costs that are incurred by the owner of the building. An industrial hygienist (IH) ensures that asbestos abatement proceeds in a manner consistent with the contract specifications and appropriate regulatory requirements. Costs for IH services include personnel, air and bulk sampling, final clearance sampling, and IH professionals or technical labor to monitor the work as it progresses.

Project monitoring through air sampling and analysis is performed throughout the project. The air samples are taken using a vacuum pump that draws air through a filter approximately 1 inch in diameter. The filter is then analyzed for total fibers using phase contrast microscopy (PCM) at a cost of about $40 per sample. The average number of samples per shift is approximately eight, obtained in the following locations:

- Two samples inside the containment
- One sample in the clean room
- One sample just outside decontamination unit
- Two HEPA filter outlet samples (one at each filter)
- One sample in offices near the abatement area
- One duplicate sample

The analytical cost per shift for project monitoring is thus:

Eight samples per 10-hour shift × $35/each = $280

Final clearance sampling is done to certify the area as being clean. A leaf blower is used to disturb the air within the containment while the air sample

Table 8.39 Laboratory Services

Analysis	UOM	Unit Cost, $/UOM
Phase Contrast Microscopy (PCM)	Each	$40
Transmission Electron Microscopy (TEM) 24-hour turnaround	Each	$600
TEM 1-week turnaround	Each	$300-$400

NOTE: The March 9, 1989, issue of Engineering News Record contains articles of interest for related references.

is obtained (aggressive air sampling). Then samples are analyzed using PCM. One sample per 5,000 square feet or a minimum of three samples is taken. PCM cannot distinguish between asbestos and nonasbestos fibers, but usually the total fiber count is so low that even if all the fibers were asbestos, the sample would pass. If the fiber count in the PCM sample is high (normally greater than 0.01 fibers/cc), then the abatement contractor may have the option to either reclean the area or request that the sample be analyzed using transmission electron microscopy (TEM). The TEM analysis can cost as much as $600 each for a 24-hour turnaround and between $300 and $400 for a one-week turnaround. Because the contractor is usually on a tight construction schedule (due to working in an operating building), he will usually opt to reclean the area so that the crews are not idle for 24 hours waiting for the results of the analysis. For estimating purposes, an assumption that TEM will be required for 2% of the samples should be adequate. Refer to Table 8.39 for a summary of costs for laboratory services.

An IH technician also is required to monitor the work during each shift. The burdened cost of the IH technician averages $45 per hour. A senior IH technician will be required half of the project time. The burdened cost for the senior IH technician is $62.50 per hour. A project director is required for 20% of the Senior IH technicians' time and averages $90 per hour.

8.2.9 Other Considerations

One observation recently commented upon stated that a study of the cost impact of cumulative regulations (OSHA, Mine Safety, Nuclear Regulatory, DOE, etc.) increased baseline costs by a multiple of three. Although this may not necessarily be representative of all industry, it highlights a concern. We must have an awareness of requirements and a corresponding knowledge of costs associated with compliance to deter and avoid serious cost and schedule overruns, and to mitigate legal liabilities.

A recent project for a major refinery in the Delaware River area required tie-ins to existing electrical grid systems where file copies of as-built drawings had not been updated with subsequent plant modifications. Excessive engineering manhours were expended in research and rework of design. Another situation was encountered of environmental concern, and one that had not been anticipated in the owner's appropriation of costs, where hydrocarbons were discovered during excavations for underground electrical ductbanks. Precautions taken for monitoring, subcontractor halts and standby time, and alternative construction methods added $400K to the project cost. Had a test for soils contamination been performed during the scope definition phase, much of the "surprise" could have been eliminated.

APPENDIX A
Acronyms
and Abbreviations

A	Approval
AAAP	Advanced acquisition assistance plan
ACM	Asbestos-containing materials
ACWP	Actual cost of work performed
AD	Associate director
ADC	Authorized derivative classifier
ADM	Arrow diagram method
ADP	Automatic data processing
ADPE	Automatic data processing equipment
ADS	Activity data sheet
AE	Acquisition executive
A-E	Architect-engineer
AEA	Atomic Energy Act
AEC	Atomic Energy Commission
AFA	Authorization for acceptance
AFA	Approval for final acceptance
AFP	Approved financial plan
AHERA	Asbestos Hazardous Emergency Response Act
AIP	Agreement in principle
AL	Albuquerque Field Office
ALARA	As low as reasonably achievable
AMO	Assistant manager for operations
AME	Assistant manager for environmental management
AMP	Assistant manager for projects

AMT	Assistant manager for technical support
ANSI	American National Standards Institute
AOC	Area of concern
AOP	Annual operational plan
AP	Activity package
APV	Accounts payable voucher
ARAR	Applicable or relevant and appropriate requirements
ASME	American Society of Mechanical Engineers
ASNE	American Society of Nuclear Engineers
AST	Above-ground storage tanks
ATP	Acceptance test procedures
ATSDR	Agency for toxic substances and disease registry
AUW	Authorized unpriced work
B&R	Budget and reporting
BA	Budget authorization
B/A	Budget authority
BAC	Budget at completion
BACT	Best available control technology
BAFO	Best and final offer
BAT	Best available technology
BCCB	Baseline change control board
BCO	Bid control office
BCP	Baseline change package
BCP	Baseline change proposal
BCY	Bank cubic yards
BCWP	Budget cost of work performed
BCWS	Budget cost of work scheduled
BMP	Best management practices
BO	Budget outlay or expenditure
BRC	Budget review committee
BS&F	Business service and finance
BSG	Business strategy group
BSRI	Bechtel Savannah River Inc.
BTC	Business and technology center
BY	Budget year
CA	Corrective action
CA	Cost account
CAA	Clean Air Act
CAC	Cost at completion
CAD	Computer-aided design and drafting
CAM	Cost account manager
CCN	Contract change notice
CAP	Cost account plan
CAPCA	Closure and post closure activities

CCB	Change control board
CCE	Certified cost engineer
CCY	Compacted cubic yards
CD	Conceptual design
CDC	Center for disease control
CDR	Conceptual design report
CDS	Construction data system
CE	Civil engineer
CENRTC	Capitol equipment not related to construction
CEQ	Council on Environmental Quality
CER	Cost estimating relationship
CERCLA	Comprehensive Environmental Response, Compensation and Liability Act
CF	Cubic feet
CFR	Code of Federal Regulations
CFY	Current fiscal year
CH	Chicago Field Office
CHE	Chemical engineer
CICA	Competition in contracting act
CIL	Critical items list
CLP	Contract laboratory program
CLS	Consolidated labor system
CM	Construction manager
CMI	Corrective measures implementation
CMI	Corrective measures investigation
CMP	Configuration management plan
CMP	Corrugated metal pipe
CMP	RL Office of Compliance
CMS	Corrective measures study
COE	U.S. Army Corps of Engineers
COR	Contracting officer representative
CPAF	Cost plus award fee
CPI	Cost performance index
CPM	Critical path method
CPR	Cost performance report
CPS	Conceptual project schedule
CR	Change request
CR	Community relations
CRP	Community relations plan
CRRFI	Clinch River RCRA Facility Investigation
C/SCS	Cost/schedule control system
C/SCSC	Cost/schedule control system criteria
CSMIS	Cost and schedule management information system
CTC	Cost to complete

CV	Cost variance
CVI	Certified vendor information
CWA	Clean Water Act
CX	Categorical exclusion
CYWP	Current year work plan
D&D	Decontamination and decommissioning
DEAR	Department of Energy acquisition regulation
DE/EM MON	Defense Programs EM Memo of Understanding
DEIS	Draft environmental impact statement
DIES	Design information exchange system
DISCAS	Departmental Integrated Standardized Core Accounting System
DNFSB	Defense Nuclear Facility Safety Board
DOD	Department of Defense
DOE	Department of Energy
DOE/GJPO	Department of Energy Grand Junction Program Office
DOE/HQ	Department of Energy/Headquarters
DOE/ID	Department of Energy/Idaho Field Office
DOI	Department of Interior
DOL	Department of Labor
DOT	Department of Transportation
DP	Office of Defense Programs
DQO	Data quality objectives
DWPF	Defense waste processing facility
EA	Environmental assessment
EA	Each
EAC	Environmental Advisory Committee
EAC	Estimate at completion
E/C	Engineer/constructor contractor
ECA	Engineering cost analysis
ECN	Engineering change notice
ED&I	Engineering design and investigation
EE	Electrical engineer
EE	Environmental evaluation
EEO	Equal employment opportunity
EFPC	East Fork Poplar Creek (OR-Y12)
EH	Office of Environmental Safety and Health
EIS	Environmental impact statement
EM	Office of Environmental Restoration and Waste Management
EMTC	Environmental Management Technical Center (AL-5)
EM-1	Director of the Office of Environmental Restoration and Waste Management
EM-10	Office of Planning and Resource Management
EM-20	Office of Quality Assurance and Quality Control
EM-30	Office of Waste Operations (OWO)

EM-40	Office of Environmental Restoration (OER)
EM-50	Office of Technology Development (OTD)
ENERGY SYSTEMS	Martin Marietta Energy Systems, Inc.
EP	Extraction procedure
EPA	Environmental Protection Agency
EPCRA	Emergency Planning and Community Right-to-Know Act
EQAB	Environmental Quality Advisory Board
ER	Office of Energy Research
ER	Environmental restoration
ERA	Expedited response action
ERDA	Energy Research and Development Agency
ERP	Environmental Restoration Program
ESAAB	Energy Systems Acquisition Advisory Board
ES&H	Environmental, safety and health
ETC	Estimate to complete
4 Cs	Construction, completion and cost closing statement
FA	Force account
FDC	Functional design criteria
FFA	Federal facility agreement
FFCA	Federal Facility Compliance Agreement
FIFRA	Federal Insecticide, Fungicide, and Rodenticide Act
FINPLAN	Financial plan
FIS	Financial information system
FIS	Richland Financial Information System
FMD	Richland Financial Management Division
FML	Flexible membrane liner
FMPR	Feed materials production center
FMPC	Feed materials production control
FONSI	Finding of no significant impact
FP	Fixed price
FPM	Federal Personnel Manual
FPMR	Federal Property Management Regulation
FS	Feasibility study
FSAR	Final safety analysis report
FSP	Field sampling plan
FTE	Full-time employees
FUSRAP	Formerly Utilized Sites Remedial Action Program
FY	Fiscal year
FYP	Five-year plan
G&A	General and administrative expense
GC	Office of General Council
GCD	General containment disposal
GCEP	Gas Centrifuge Expansion Program (ORNL)

GJ	Grand Junction
GN	Geonet
GOCO	Government-owned, Contractor-operated
GPM	Gallons per minute
GPP	General Plant Project
GT	Geotextile
GWPS	Ground water protection strategy
H&S	Health and safety
HAZWRAP	Hazardous Waste Remedial Actions Program
HDPE	High-density polyethylene
HEPE	High-efficiency particulate air
HHRA	Human health risk assessment
HLLW	High level liquid waste
HMTA	Hazardous Materials Transportation Act
HNu	(Brand Name) Photoionization Detector
HPS	Hanford Plant Standards
HQ	Headquarters
HRS	Hazard ranking system
HSDP	Hanford Site Development Plan
HSP	Health and safety plan
HSWA	Hazardous and Solid Waste Amendments (to RCRA)
HVAC	Heating, ventilating, and air conditioning
HWV	RL Hanford Waste Vitrification Project Division
I	Information
IH	Industrial hygienist
IAG	Interagency agreement
IAW	In accordance with
IBARS	Integrated budget accounting and reporting system
ICE	Independent cost estimate
ID	Idaho Field Office
IHSS	Individual hazardous substance sites
IM	Interim measures
INEL	Idaho National Engineering Laboratory
IPS	Integrated project schedule
IRA	Interim remedial action
IRM	Interim remedial measure
IT	IT Corporation
ISV	In-site vitrification
I-T-D	Inception-to-date
IVC	Independent verification contract
JEV	Journal entry voucher
K	Thousands
K-25	Oak Ridge Gaseous Diffusion Plant
KD	Key decision

LCC	Life cycle costs
LCRS	Leachate collection and removal system
LDCRS	Leak detection, collection, and removal system
LEL	Lower explosive limit
LF	Linear feet
LI	Line item project
LLW	Low level waste
LOE	Level of effort
M	Meters
M	Thousands
M&O	Management and operation contractor
M&O	Management and operation
MCL	Milestone control log
MCL	Maximum contaminant level
MCLG	Maximum contaminant level goal
ME	Mechanical engineer
MED	Manhattan Engineer District
MM	Millions
MMES	Martin Marietta Energy Systems
MOU	Memorandum of understanding
MP	Management plan
MP	Major project
MPR	Management policies and requirements
MR	Management reserve
MRAP	Monticello Remedial Action Project
MRM	Management review meeting
MSA	Major systems acquisition
MSDD	Management system design description
MSDS	Material safety data sheet
M-T-D	Month-to-date
MTF	Memorandum to file
MVP	Monticello Vicinity Properties
MW	Mixed waste
MWMF	Mixed waste management facility
MWTA	Medical Waste Tracking Act
N	Standard Penetration Value in Number of Blows of the Hammer per Foot of Penetration
NAS	National Academy of Science
NASA	National Aeronautics and Space Administration
NCP	National Contingency Plan
NE	Office of Nuclear Energy
NCR	Nonconformance report
NEPA	National Environmental Policy Act
NHP	New Hope Pond

NLT	Not later than
NOA	Notice of authorization
NOI	Notice of intent
NPDES	National Pollutant Discharge Eliminating System
NPL	National Priorities List
NQA-1	Nuclear Quality Assurance-1
NRC	Nuclear Regulatory Commission
NRT	National Response Team
NRWTP	Non-Radiological Waste Treatment Project
NS	Office of Nuclear Safety
NTS	Nevada Test Site
NV	Nevada Field Office
OAC	Official acceptance of construction
OBS	Organizational breakdown structure
OC	Operating contractor (Q&E and R&D contractors)
OCC	Office of Chief Counsel
ODC	Other direct cost
O&E	Operations and Engineering Contractor
OCRWM	Office of Civilian Radioactive Waste Management
OERR	Office of Emergency and Remedial Response
O&M	Operation and maintenance
OMB	Office of Management and Budget
O/O	Owner/operator
OR	Oak Ridge Field Office
ORAU	Oak Ridge Associated Universities
ORGDP	Oak Ridge Gaseous Diffusion Plant (K-25 Site)
ORNl	Oak Ridge National Laboratory
ORO	Oak Ridge Operations
ORR	Operational readiness review
OSHA	Occupational Safety and Health Administration
OSWER	Office of Solid Waste and Emergency Response
OTD	Office of Technology Development (EM-50)
OTP	Operations test procedures
OU	Operable unit
OVA	Organic vapor analyzer
OWO	Office of Waste Operations (EM-30)
PA	Preliminary assessment
PA	Project authorization
P&A	Plug and abandon
PA/SI	Preliminary assessment/site inspection
PACE	Plant and capitol equipment (budget)
PBB	Project budget base
PB&C	Planning, budgeting and control
PBD	RL plans and budget division

PCB	Polychlorinated biphenyl
PCE	Project control engineer
PCM	Phase contrast microscopy
PCR	Project cost report
PD	Program decision
PDS	Project data sheet (schedule 44)
PE	Professional engineer
PEIS	Preliminary environmental impact study
PFSR	Project funds status report
PHS	Public Health Service
PIP	Project inspection plan
PLM	Polarized light microscopy
PM-10	Particulate matter less than 10 microns
PMB	Performance measurement baseline
PMD	RL Project Management Division
PMP	Program management plan
PMPR	Program management policies and requirements
PMS	Program management system
POTW	Publicly-owned treatment works
PP	Proposed plan
PPB	Parts per billion
PPE	Personal protective equipment
PPM	Parts per million
PRO	RL Procurement Division
PRP	Potentially responsible party
PSAR	Preliminary safety analysis report
PSCS	Preliminary site characterization summary
PSD	Prevention of significant determination
PSD	Program summary document
PSE	Preliminary safety evaluation
PSO	Program secretarial officer
PSWBS	Program summary waste breakdown structure
PTS	Project tracking system
PUF	Polyurethane foam
PU/U	Plutonium/uranium
PWA	Process waste assessment
PWE	Present working estimate
QA	Quality assurance
QAPD	Quality assurance program description
QAPP	Quality assurance project plan
QC	Quality control
QPP	Quality program plan
R	Review
R&D	Research and development

RA	Remedial action
RAA	Remedial action agreement
RAM	Responsibility assignment matrix
RAP	Remedial action program
RCD	Reference conceptual design
RCRA	Resource Conservation and Recovery Act
RD	Remedial design
RD&D	Research, development, and demonstration
RD/RA	Remedial design/remedial action
RDDT&E	Research, development, demonstration, testing, and evaluation
REA	Radiologic and engineering assessment
REC	Request for engineering change
RFA	RCRA Facility Assessment
RFA	Request for funds authorization
RFD	Reference dose
RFI	RCRA Facility Investigation
RI	Remedial investigation
RI/FS	Remedial investigation/feasibility study
RL	DOE Field Office, Richland
RLIP	DOE Field Office, Richland Implementing Procedure
ROD	Record of decision
ROD	Record of deviation
ROM	Rough order of magnitude
RPA	Request for project authorization
RPAM	Request for project authorization modification
RPM	Reactive plum model
RPM	Remedial project manager (CERCLA)
RPM	Revolutions per minute
RWMS	Radioactive waste management site
S&M	Surveillance and monitoring
SAP	Sampling and analysis plan
SARA	Superfund Amendments and Reauthorization Act
SAS	RL Safeguards and Security Division
SBA	Small Business Act, Section 8(a)
SDWA	Safe Drinking Water Act
SEEP	Groundwater into surface water
SEMP	System engineering management plan
SER	Site evaluation report
SFDS	Short form data sheet
SFMP	Surplus facilities management program
SI	Site inspection
SID	RL Site Infrastructure Division
SITE	Superfund innovative technology evaluation
SOW	Statement of work

SPDES	State pollutant discharge elimination system
SRD	Supplemental design requirements document
SP	Small projects
SR	Savannah River Field Office
SRL	Savannah River Laboratory
SRS	Savannah River Site
SSDW	Subproject scope definition worksheet
SSP	Site-specific plans
SST	Single-shelled tanks
SV	Schedule variance
SWMU	Solid waste management unit
SWSA	Solid waste storage area
T	Ton
TAC	Technical assistance contractor
TAG	Technical assistant grant
TBC	To be considered
TCE	Trichloroethylene/Tetrachloroethylene
TCL	Target compound list
TDHE	Tennessee Department of Health and Environment
TEC	Total estimated cost
TECC	Total estimated construction cost
TEM	Transmission electron microscopy
TLD	Toxic lethal dose
TM	Technical monitor
TNT	Trinitrotoluene
TPC	Total project cost
TPCE	Total project cost estimate
TPD	Tons per day
TPY	Tons per year
TRU	Transuranic
TSCA	Toxic Substance Control Act
TSD	RL Technical Support Division
TSWMA	Tennessee Waste Management Act
T/S/D	Treatment/storage/disposal
TVA	Tennessee Valley Authority
TWC	Type of work category
UB	Undistributed budget
UCC	Union Carbide Corporation
UEFPC	Upper East Fork Poplar Creek (ORNL)
UMTRA	Uranium Mill Tailings Remodiation Act
UST	Underground storage tank
UV	Ultraviolet
VAC	Variance at completion
VAR	Variance analysis report

VCA   Vanadium Corporation of America
VE   Value engineering
VLF   Vertical linear foot
VOC   Volatile organic compounds
WA   Waste authorization
WAG   Waste area group
WBS   Work breakdown structure
WIPP   Waste isolation pilot plant
WM   Waste management operations
WQC   Water quality criteria
WSRC   Westinghouse Savannah River Company
X-10   Oak Ridge National Laboratory
Y-T-D   Year-to-date
Y-12   Oak Ridge Y-12 Plant

APPENDIX B
Glossary

acceleration Acceleration of the contract occurs when a contractor is forced to perform work in a shorter period of time than called for in the contract. Actual acceleration occurs when the owner directs that the work is to be completed earlier than required by contract. Constructive acceleration occurs when the contractor has demonstrated that a time extension is required but the owner has denied it, thereby requiring the contractor to complete the work in accordance with a schedule that has not been properly extended.

abatement Procedures to control fiber release from asbestos-containing materials. Includes removal, encapsulation, enclosure, repair, demolition, and renovation activities.

aggressive air sampling A method used to stir the air by use of a blower fan while final air testing is performed.

airlock A system for permitting ingress and egress with minimum air movement between a contaminated area and an uncontaminated area, typically consisting of two curtained doorways separated by a distance of at least 3 feet such that a person can pass through one doorway into the airlock allowing the doorway sheeting to overlap and close off the opening before proceeding through the second doorway, thereby preventing flow-through contamination.

air monitoring The process of measuring the fiber content of a known volume of air collected during a specific period of time to determine the concentration of asbestos fibers. Phase contrast microscopy (PCM) is the technique used to count fibers. If the fiber count is too high, electron microscopy (TEM) may be used for lower detectability and specific fiber identification.

alternate dispute resolution (ADR)   The process used to resolve a contract dispute as a prerequisite to litigation. ADRs typically use a third party to assist in the negotiation and resolution of the dispute. ADRs attempt to avoid the costly process of litigation.

amended water   Water to which a surfactant has been added.

arbitration   A typical dispute resolution process involving the use of an arbiter. Arbitration can be either binding or nonbinding, depending on the provisions of the contract.

asbestos   Asbestiform varieties of serpentine (chrysotile), rie becktie (crocidolite), cumingtonite-grunerite (amosite), anthophyllite, actinolite, and tremolite.

asbestos abatement worker   Laborers trained in asbestos abatement. This classification of workers must be used when abating pipes, etc., on mechanical equipment.

asbestos-containing material (ACM)   Material composed of asbestos of any type in an amount greater than 1% by weight, either alone or mixed with other fibrous or nonfibrous materials.

asbestos-containing waste material   Asbestos-contaminated objects or asbestos-containing material requiring disposal.

as-built schedule   Schedule prepared at the end of the project (or after) indicating actual completion dates and durations for all activities. The as-built schedule reflects any logic changes, scope additions, resequencing, and delays experienced during the project.

as-planned schedule   Schedule, normally made by the contractor at the start of the project. As-planned schedules typically correspond with the contract times, reflect no delays, and describe the initial sequence of work as well as expected durations for all planned activities.

action level   As defined by OSHA, 0.1 fiber per cubic centimeter of air (f/cc), averaged over an 8-hour period. If this level is exceeded, employers must begin compliance activities such as air monitoring, employee training, and medical surveillance.

authorized visitor   The building owner (and any designated representatives) or any representative of a regulatory or other agency having jurisdiction over the project. The contractor should keep two respirators and extra protective clothing in the clean room for use by authorized visitors.

bonds   Payment and performance bonds are an additional way to ensure compliance with an asbestos abatement contractor.

bulk sampling   A technique used to collect samples of suspect materials such as fireproofing, pipe lagging, boiler insulation, and acoustical spray. Bulk sampling is usually conducted during the building survey and provides data for decisions on control measures. Settled dust, wipe, and tape sampling are all bulk sampling methods.

certified industrial hygienist (CIH)   An industrial hygienist certified in comprehensive practice by the American Board of Industrial Hygiene. His/her duties include the supervision of air sampling and evaluation of results. Such a person shall not be affiliated in any way other than through this contract with the contractor performing the abatement work.

changed conditions   Also known as *differing site conditions.*

changes   A process by which the contract is modified to add, delete, or revise the scope of work and adjustments made to the contract price and contract times.

claims   Formalization of a dispute. Contractor claims relate to situations where the contractor gives written notice that a condition exists that requires an adjustment in the contract time and/or cost and the owner fails to recognize it by initiating a change order or modification. Owner claims refer to perceived failure of the contractor to perform work as specified in the contract.

clean room   An uncontaminated area or room that is part of the worker decontamination enclosure system with provisions for storage of worker's street clothes and clean protective equipment.

claims avoidance reviews   Systematic reviews of design and contract documents with a focus of identifying and correcting areas susceptible to changes, claims, and disputes.

composite crew rate   A rate consisting of a supervisor and three skilled/unskilled laborers and/or asbestos abatement workers, depending on the abatement work to be performed.

contingent pricing   Unit prices in the contract, additional to the fixed price elements, which can be used in lieu of or to supplement the fixed price elements of the contract as the work develops and actual quantities become known.

consumables and small equipment   Items used during abatement (e.g., knifes, duct tape, buckets, hand sprayers, boots, and respirators).

containment   A work area that has been sealed, plasticized, and equipped with a decontamination or enclosure system.

contract award   A date by which the owner states that he/she intends to enter into a contract with the successful bidder. The written notice of this intention is called a *notice to award.*

contract management   Aspects of project management that relate to the daily interaction between the owner and contractor. This interaction includes all specific coordination roles between these parties as well as the administration of contract terms and conditions (such as payment, change orders, directions, etc.).

contract milestones   Dates established by the contract for completion and other important milestones in the project. Often contract milestones are used in conjunction with liquidated damages clauses.

contract price   Value of the work as established by the contract provisions. In lump sum contracts the contract price is firmly established. In other types of contracts the contract price is determined by quantity surveys or audits of the contractor's costs.

contract times   Period of time, normally the number of days from the notice to proceed to the completion of the work or a specified portion of the work. Typically, there are two times specified in most contracts: the number of days to substantial completion and final completion.

curtained doorway A device to allow ingress and egress from one room to another while permitting minimal air movement between the rooms, typically constructed by placing two overlapping sheets of plastic over an existing or temporarily framed doorway, securing each along the top of the doorway, securing the vertical edge of one sheet along one of the vertical sides of the doorway, and securing the vertical edge of the other sheet along the opposite vertical side of the doorway. Other effective designs are permissible.

decontamination enclosure system (DECON) A series of connected rooms from the clean room to the containment that are separated from each other by air locks, for the decontamination of workers and equipment.

demoliton The wrecking or taking out of any structure to gain access for abatement.

deck Non-load-bearing ceilings, such as corrugated steel, concrete, or plywood.

differing site conditions Contract situations where the contractor seeks monetary relief when site conditions differ materially from those expected in the contract. Type I differing site conditions refer to subsurface or latent physical conditions at the site that differ materially from those indicated in the contract. Type II differing site conditions refer to unknown physical conditions of an unusual nature at the site that differ materially from those ordinarily encountered in the type of work described in the contract.

disputes Disagreement about the interpretation of the contract, including the recognition and pricing of changes.

encapsulant A liquid material that can be applied to asbestos-containing material. It controls the possible release of asbestos fibers from the material either by creating a membrane over the surface (bridging encapsulant) or by penetrating into the material and binding its components together (penetrating encapsulant).

encapsulation The application of an encapsulant to asbestos-containing materials to control the release of asbestos fibers into the air. This application is also used after the final visual inspections and before final air testing begins.

equipment decontamination enclosure system That portion of a decontamination enclosure system designed for controlled transfer of materials and equipment into or out of the work area, typically consisting of a washroom and a holding area.

equipment room A contaminated area or room that is part of the worker decontamination enclosure system with provisions for storage of contaminated clothing and equipment.

facility Any industrial, commercial, or industrial structure, installation, or building.

facility component Any pipe, duct, boiler, tank, reactor, turbine, or furnace at or in a facility or any structural member of a facility.

fast tracking Project execution strategy whereby normal sequence of design and/or construction activities are overlapped. Projects using this strategy are termed fast-track projects.

final air   When all the asbestos-containing material (ACM) has been removed from a containment and the air samples taken using aggressive air sampling techniques are less than the exposure level designated (e.g., PEL or action level).

final completion   Contract milestone whereby all work specified in the contract has been satisfactorily completed.

fixed object   A piece of equipment or furniture that cannot be removed from the work area.

float   In critical path method (CPM) scheduling, the amount of slack time between a noncritical activity and the critical path. In contract management, the amount of time between the contractor's forecast for completion and the contract times specified in the contract.

float suppression   Reducing the amount of float time in a CPM schedule through clever network techniques such as preferential sequencing, unusual use of lead/lag restraints, and/or inflated durations for activities.

force majeure   Clauses in contracts that allow time extensions for acts of God or natural disasters.

friable asbestos   Asbestos-containing material that can be crumbled to dust, when dry, under hand pressure.

glovebag technique   A method with limited applications for removing small amounts of friable asbestos-containing material from HVAC ducts, short piping runs, valves, joints, elbows, and other nonplanar surfaces in a noncontained work area. The glovebag assembly is a manufactured or fabricated device consisting of a glovebag (typically constructed of 6 mil transparent polyethylene or polyvinylchloride plastic), two inward projecting long sleeves, an internal tool pouch, and an attached, labeled receptacle for asbestos waste. The glovebag is constructed and installed in such a manner that it surrounds the object or material to be removed and contains all asbestos fibers released during the process. All workers who are permitted to use the glovebag technique must be highly trained, experienced, and skilled in this method.

HVAC   Heating, ventilation, and air conditioning system.

HEPA filter   A high-efficiency particulate air filter capable of removing particles >0.3 microns in diameter with 99.97% efficiency. HEPA filters are used in exhaust ventilation systems.

HEPA vacuum   A vacuum system equipped with HEPA filtration.

homogeneous area   Areas that may be grouped together for abatement purposes because of their similarity in work elements.

industrial hygienist (IH)   A hygienist trained to work and record asbestos abatement activities, perform air monitoring, polyvisual, and final inspection, and final air testing.

inspections   As performed by the IH; (1) initial polyvisual inspection of the two layers of plastic forming the containment walls and/or floor to ensure that there are no holes or tears in the plastic enclosure, (2) a final inspection to ensure no visible asbestos-containing material (including cracks and crevices) may be seen before a final air test is performed.

insurance Usually "claims made" insurance, which means that the insurance carrier is only liable to provide coverage if a claim is made during the policy period.

liquidated damages Contractual provisions under which the parties agree that unexcused delays will cost the contractor specified sums of money; Normally expressed as a cost per calendar day.

lump sum contract A type of contract in which the cost of the work to be performed is shown as an all-inclusive price; also known as a *fixed-price contract.*

master schedule Broad-base schedule that shows all facets of the project. master schedules are commonly used in projects with multiple prime contractors to provide overall schedule reporting.

mediation A common alternate dispute resolution (ADR) technique in which a "third-party" mediator creates a forum to negotiate.

movable object A piece of equipment or furniture that can be removed from the work area.

negative pressure ventilation system A portable exhaust system equipped with HEPA filtration and capable of maintaining a constant, low-velocity air flow into contaminated areas from adjacent uncontaminated areas. These units provide four air changes per hour.

notice to proceed The written notice issued by the owner to the contractor authorizing the contractor to proceed with the work described in the contract. Generally considered the start of the contract for compliance with the contract times requirements.

permissible exposure limit (PEL) As defined by OSHA, 0.2 fiber per centimeter of air (f/cc), over an 8-hour period.

phase contrast microscopy (PCM) A sample analysis technique using a light microscope equipped to provide enhanced contrast between the fibers and the background. The method does not distinguish between fiber types; however, it does count fibers that are longer than 5 micrometers and wider than about 0.25 microneters.

plasticize To cover floors and walls with plastic sheeting as herein specified. Typically, two layers of 6-mil polyethylene or polyvinylchloride plastic sheet on floors, walls, and objects in the containment.

polarized light microscopy (PLM) A bulk sample analysis method consisting of using a light microscope equipped with polarizing filters. Identification of asbestos is based on the determination of optical properties displayed when the sample is treated with various dispersion staining liquids.

protective clothing Disposable clothing and filters for respirators as required when abating.

record schedule A contemporaneous schedule used to document the actual starts and completions of activities in a project. If done correctly, the final version of the record schedule can also serve as the as-built schedule.

reimbursable cost contract A type of contract where the cost of the work is determined by actual cost incurred by the contractor. Profit and overhead costs of the contractor can be a fixed amount or a percentage of the actual costs. Also commonly termed as a *time-and-materials* contract.

requests for information (RFIs) A formal request by the contractor to clarify the technical aspects of the contract.

scanning electron microscopy (SEM) A sample analysis method that directs an electron beam onto the sample surface and collects those beams that are reflected. This method can identify fibers greater than 0.05 micrometers in length.

shower room A room between the clean room and the equipment room in the worker decontamination enclosure with running water controllable at a tap and suitably arranged for complete showing during decontamination.

structural member Any load-supporting member of a facility, such as steel beams and precast concrete beams.

substantial completion A contract milestone for completing the work (or a specified portion of the work). At substantial completion the owner has the ability and option to occupy and use the new facilities for their intended purpose. Only minor non-essential work items, such as punchlist items or landscaping, remain after substantial completion.

surfactant A chemical wetting agent added to water to improve penetration.

unit price contract A type of contract in which the cost of the work is shown as a schedule of unit prices to be used against a measured set of quantities.

value engineering A technical review process in which changes to the specified facilities are considered in an effort to reduce the total cost of the project. In the case of a contractor-initiated request for value engineering changes, many contracts allow a sharing of the recovered savings due to the new idea.

waste transfer airlock A decontamination system used for transferring containerized waste from the inside to the outside of the work area.

wet wipe The process of eliminating asbestos contamination from the building surfaces and objects by using cloths, mops, or other cleaning tools that have been dampened with water, or amended water, and afterwards thoroughly decontaminated or disposed of as asbestos-contaminated waste. This process is used for pre-cleaning through final detail cleaning.

work area Designated rooms, spaces, or areas of the project in which asbestos abatement actions are to be undertaken or which may become contaminated as a result of such abatement actions. A noncontained work area is an isolated or controlled-access work area that has not been plasticized nor equipped with a decontamination enclosure system (e.g., an area where glovebags are performed).

APPENDIX C
Regulatory Agency References

William A. Zbitnoff

Table C.1 Environmental Protection Agency Addresses

EPA Regions	States Within Region	EPA Regional Addresses
Region I	Connecticut, Maine, Massachusetts, New Hampshire, Rhode Island, and Vermont	JFK Federal Building, Room 2203 Boston, Massachusetts 02203
Region II	New Jersey, New York, and Puerto Rico	26 Federal Plaza New York, New York 10278
Region III	Delaware, District of Columbia, Maryland, Pennsylvania, Virginia, and West Virginia	841 Chestnut Street Philadelphia, Pennsylvania 19107
Region IV	Alabama, Florida, Georgia, Kentucky, Mississippi, North Carolina, South Carolina, and Tennessee	345 Courtland Street N.E. Atlanta, Georgia 30364
Region V	Illinois, Indiana, Michigan, Minnesota, Ohio, and Wisconsin	230 South Dearborn Street Chicago, Illinois 60604
Region VI	Arkansas, Louisiana, New Mexico, Oklahoma, and Texas	1445 Ross Avenue, Suite 1200 Dallas, Texas 75202
Region VII	Iowa, Kansas, Missouri, and Nebraska	726 Minnesota Avenue Kansas City, Kansas 66101

Table C.1   (*continued*)

EPA Regions	States Within Region	EPA Regional Addresses
Region VIII	Colorado, Montana, North Dakota, South Dakota, Utah, and Wyoming	One Denver Place 999 18th Street, Suite 500 Denver, Colorado 80202
Region IX	Arizona, California, Guam, Hawaii, and Nevada	215 Fremont Street San Francisco, California 94105
Region X	Alaska, Idaho, Oregon, and Washington	1200 Sixth Avenue Seattle, Washington 98101

STATE HAZARDOUS WASTE MANAGEMENT AGENCIES

Alabama

Alabama Department of Environmental Management
1751 Federal Drive
Montgomery, Alabama 36130

Alaska

Alaska Department of Environ-
mental Conservation
Pouch O
Juneau, Alaska 99811

Arizona

Arizona Department of Health
Services
2005 North Central, Room 301
Phoenix, Arizona 85004

Arkansas

Solid and Hazardous Waste
Division
Arkansas Department of Pollution
Control and Ecology
P.O. Box 9583
Little Rock, Arkansas 72219

California

Department of Health Services
714 P Street
Sacramento, California 95814

Colorado

Waste Management Division
Colorado Department of Health
4210 East 11th Avenue
Denver, Colorado 80220

Connecticut

Hazardous Materials Management
Unit
Connecticut Department of En-
vironmental Protection
165 Capitol Avenue
Hartford, Connecticut 06106

Delaware

Delaware Department of Natural
Resources and Environmental
Control
89 Kings Highway
P.O. Box 1401
Dover, Delaware 19901

District of Columbia

Department of Consumer and
  Regulatory Affairs
Pesticides and Hazardous Waste
  Branch
5010 Overlook Avenue S.W.,
  Room 114
Washington, DC 20032

Florida

Solid and Hazardous Waste
Florida Department of Environmental Regulation
2600 Blair Stone Road
Tallahassee, Florida, 32301

Georgia

Georgia Department of National
  Resources
Land Protection Branch
270 Washington Street S.W.,
  Room 723
Atlanta, Georgia 30334

Hawaii

Hawaii Department of Health
Environmental Protection and
  Health Services Division
Noise and Radiation Branch
P.O. Box 3378
Honolulu, Hawaii 96810

Idaho

Idaho Department of Health and
  Welfare
450 West State Street
Boise, Idaho 83720

Illinois

Division of Land Pollution Control
Illinois Environmental Protection
  Agency
2200 Churchill Road
Springfield, Illinois 62706

Indiana

Hazardous Waste Management
  Branch
Division of Land Pollution
  Control
Indiana State Board of Health
1330 West Michigan Street
Indianapolis, Indiana 46206

Iowa

Iowa Department of Water, Air
  and Waste Management
900 East Grand
Des Moines, Iowa 50319

Kansas

Kansas Department of Health and
  Environment
Forbes Field
Topeka, Kansas 66620

Kentucky

Kentucky Department of Environmental Protection
18 Reilly Road
Frankfort, Kentucky 40601

Louisiana

Office of Solid and Hazardous
  Waste
Louisiana Department of Environ-
  mental Quality
P.O. Box 94307
Baton Rouge, Louisiana 70804

Maine

Bureau of Oil and Hazardous
  Materials Control
Maine Department of Environ-
  mental Protection
State House—Station 17
Augusta, Maine 04333

Maryland

Maryland Office of Environmental
Programs
201 West Preston Street
Baltimore, Maryland 21201

Massachusetts

Division of Solid and Hazardous
Waste
Massachusetts Department of
Environmental Quality
Engineering
One Winter Street, 5th Floor
Boston, Massachusetts 02108

Michigan

Hazardous Waste Division
Michigan Department of Natural
Resources
P.O. Box 30038
Lansing, Michigan 48909

Minnesota

Minnesota Pollution Control
Agency
1935 West Country Road B2
Roseville, Minnesota 55113

Mississippi

Mississippi Department of Natural
Resources
Bureau of Pollution Control
Division of Solid Waste Manage-
ment
P.O. Box 10385
Jackson, Mississippi 39209

Missouri

Missouri Department of Natural
Resources
Division of Environmental Quality
P.O. Box 1368
Jefferson City, Missouri 65102

Montana

Montana Department of Health
and Environmental Services
Solid and Hazardous Waste Man-
agement Bureau
Cogswell Building, Room
B-201
Helena, Montana 59620

Nebraska

Nebraska Department of Environ-
mental Control
Hazardous Waste Management
Section
P.O. Box 94877
Statehouse Station
301 Centennial Mall South
Lincoln, Nebraska 68509

Nevada

Nevada Department of Conserva-
tion and Natural Resources
Division of Environmental
Protection
Capitol Complex
Carson City, Nevada 89710

New Hampshire

New Hampshire Division of Public
Health Services
Office of Waste Management
Health and Welfare Building
Hazen Drive
Concord, New Hampshire 03301

New Jersey

Division of Waste Manage-
ment
New Jersey Department of Environmental Protection
P.O. Box CN028
32 East Hanover Street
Trenton, New Jersey 08625

New Mexico

Hazardous Waste Section
New Mexico Environmental
Enforcement Division
P.O. Box 968
Santa Fe, New Mexico 87504

New York

Division of Solid and Hazardous
Waste
New York Department of Environ-
mental Conservation
50 Wolf Road
Albany, New York 12233

North Carolina

North Carolina Department of
Human Resources
Solid and Hazardous Waste Man-
agement Branch
P.O. Box 2091
Raleigh, North Carolina 27602

North Dakota

Division of Hazardous Waste
Management and Special Studies
North Dakota State Department
of Health
1200 Missouri Avenue, Room 302
Bismarck, North Dakota 58501

Ohio

Division of Solid and Hazardous
Waste Management
Ohio Environmental Protection
Agency
361 Broad Street
Columbus, Ohio 43215

Oklahoma

Waste Management Section
Oklahoma State Department of
Health
P.O. Box 53531
Oklahoma City, Oklahoma 73152

Oregon

Oregon Department of Environ-
mental Quality
Hazardous and Solid Waste
Division
522 S.W. 5th Avenue
Portland, Oregon 97207

Pennsylvania

Pennsylvania Department of En-
vironmental Resources
P.O. Box 2063
Harrisburg, Pennsylvania 17120

Rhode Island

Rhode Island Department of Environmental Management
204 Cannon Building
75 Davis Street
Providence, Rhode Island 02908

South Carolina

South Carolina Department of
Health and Environmental
Control
J. Marion Sims Building
2600 Bull Street
Columbia, South Carolina 29201

South Dakota

South Dakota Department of
Water and Natural Resources
Foss Building
523 East Capitol
Pierre, South Dakota 57501

Tennessee

Tennessee Division of Solid
Waste Management
Customs House, 4th Floor
701 Broadway
Nashville, Tennessee 37219

Texas

Industrial and Industrial Service
  Facilities
Industrial Solid Waste Section
Texas Department of Water
  Resources
P.O. Box 13987, Capitol Station
Austin, Texas 78711

Utah

Utah Division of Environmental
  Health
P.O. Box 45500
State Office Building, Room 4321
Salt Lake City, Utah 84145

Vermont

Vermont Agency of Environmen-
  tal Conservation
Air and Solid Waste Division
State Office Building
Montpelier, Vermont 05602

Virginia

Virginia Department of Health
Division of Solid and Hazardous
  Waste Management
Monroe Building, 11th Floor
101 North 14th Street
Richmond, Virginia 23219

Washington

Washington Department of
  Ecology
Office of Hazardous Substances
Mail Stop PV-11
Olympia, Washington 98504

West Virginia

Solid and Hazardous Waste/
  Ground Water Branch
West Virginia Department of
  Natural Resources
Division of Water Resources
1201 Breenbriar Street
Charleston, West Virginia 25311

Wisconsin

Wisconsin Department of Natural
  Resources
Bureau of Solid Waste Manage-
  ment
P.O. Box 7921
Madison, Wisconsin 53707

Wyoming

Water Quality Division
Wyoming Department of Environ-
  mental Quality
Herschler Building, 4th Floor West
122 West 25th Street
Cheyenne, Wyoming 82002

Index

AACE (American Association of Cost Engineers), 2, 132–133, 141, 154–155
AACE International publications, 2, 132–133, 141, 154–155
Abandoned mine site example, 178–180
Acid Etching, 202–203
Actinolite, 209
Agency involvement, 24
Air monitoring, 223
Air pollution, 162
Allowance for change (AFC), 143
Alternate dispute resolution (ADR) options, 43
Alternative identification, 174
Alternative screening, 174
Amosite, 209
Amphibole, 209

American Association of Cost Engineers, 2, 132–133, 141, 154–155
Anthophyllite, 209
Area mixing, 191
Asbestos abatement, 207–208, 212, 223–239
  building inspection, 211–212, 225–226
  clean-up and disposal, 219–220
  contractor labor burden, overhead and profit, 235–237
  debris removal, 233–234
  decontamination, 194
  project cost estimating, 223–229
Asbestos-containing materials (ACM), 208
Asbestos Hazardous Emergency Response Act (AHERA), 210, 212

267